Practical Global Illumination with Irradiance Caching

Synthesis Lectures on Computer Graphics and Animation

Editor
Brian A. Barsky, *University of California, Berkeley*

High Fidelity Haptic Rendering
Miguel A. Otaduy, Ming C. Lin
2006

A Blossoming Development of Splines
Stephen Mann
2006

Practical Global Illumination with Irradiance Caching
Jaroslav Křivánek and Pascal Gautron

ISBN: 978-3-031-79539-8   paperback
ISBN: 978-3-031-79540-4   ebook

DOI 10.1007/978-3-031-79540-4

A Publication in the Springer series
SYNTHESIS LECTURES ON COMPUTER GRAPHICS AND ANIMATION

Lecture #10
Series Editor: Brian A. Barsky, *University of California, Berkeley*

Series ISSN
Synthesis Lectures on Computer Graphics and Animation
Print 1933-8996   Electronic 1933-9003

Practical Global Illumination with Irradiance Caching

Jaroslav Křivánek
Cornell University and Charles University, Prague

Pascal Gautron
Thomson Corporate Research, France

SYNTHESIS LECTURES ON COMPUTER GRAPHICS AND ANIMATION #10

ABSTRACT

Irradiance caching is a ray tracing-based technique for computing global illumination on diffuse surfaces. Specifically, it addresses the computation of indirect illumination bouncing off one diffuse object onto another. The sole purpose of irradiance caching is to make this computation reasonably fast. The main idea is to perform the indirect illumination sampling only at a selected set of locations in the scene, store the results in a cache, and reuse the cached value at other points through fast interpolation.

This book is for anyone interested in making a production-ready implementation of irradiance caching that reliably renders artifact-free images. Since its invention 20 years ago, the irradiance caching algorithm has been successfully used to accelerate global illumination computation in the *Radiance* lighting simulation system. Its widespread use had to wait until computers became fast enough to consider global illumination in film production rendering. Since then, its use is ubiquitous. Virtually all commercial and open-source rendering software base the global illumination computation upon irradiance caching.

Although elegant and powerful, the algorithm in its basic form often fails to produce artifact-free images. Unfortunately, practical information on implementing the algorithm is scarce. The main objective of this book is to expose the irradiance caching algorithm along with all the details and tricks upon which the success of its practical implementation is dependent. In addition, we discuss some extensions of the basic algorithm, such as a GPU implementation for interactive global illumination computation and temporal caching that exploits temporal coherence to suppress flickering in animations.

Our goal is to expose the material without being overly theoretical. However, the reader should have some basic understanding of rendering concepts, ray tracing in particular. Familiarity with global illumination is useful but not necessary to read this book.

KEYWORDS

computer graphics, realistic image synthesis, rendering, global illumination, irradiance caching, GPU

Contents

Preface

The process of generating, or *rendering*, images is a central topic of computer graphics. The key for assessing the quality of computer-generated images is their plausibility since synthetic images often have to be seamlessly integrated with real environments or have to look as if the content was real. Accurate simulation of light interacting with a scene is an important aspect in generating plausible images. This particularly involves computing global illumination that is simulating multiple reflections of light in the scene.

Lighting simulation has been an intense research area for several decades. Starting with studies on radiative transfer in the 80s, many methods have been proposed and tried, and are under continuous improvement both in the academic research and industry. However, while the main ideas are not difficult to understand, the methods for the actual computation of light transport usually present a tradeoff between speed and simplicity. Simple and robust methods are slow and faster techniques (such as irradiance caching) usually involve more complex algorithms and/or data structures. Furthermore, the limitations of the various methods are generally not extensively described in the related publications. As a result, each programmer has to find his or her own ways to uncover non-described aspects of the algorithm, and solve open issues to obtain an actually working piece of software. This systematic, repeated additional work yields an important waste of effort, and the related improvements often remain buried within the code with `/* This division by pi makes sense */` or similar comments.

To overcome this problem, this book provides extensive details on a particular method for accelerating lighting simulation, *irradiance caching*. Starting with a description of the fundamental principles of the algorithm, the book covers both theoretical and practical aspects of irradiance caching. Besides mathematical demonstrations, important aspects are discussed using pseudocode based on actual implementations of irradiance caching.

Since the invention of the irradiance caching algorithm in 1988, a number of additional research works addressed its robustness and efficiency. Also based on actual implementations, the book contains a theoretical and practical description of some relevant improvements. Solutions to some common problems encountered during implementation are also proposed.

This book is for anyone who aims at obtaining an actually working implementation of irradiance caching. Note that the reader is assumed to have a working knowledge of mathematics and computer graphics for a thorough understanding. However, we pay special attention to the accessibility of the book and provide the basic required theoretical aspects so that the book can be used as a standalone implementation reference.

WHAT IS AND WHAT IS NOT IRRADIANCE CACHING?

Irradiance caching is a ray tracing-based technique for computing global illumination on diffuse surfaces. Specifically, it addresses the computation of indirect illumination bouncing off one diffuse object onto another. The purpose of irradiance caching is to make this computation reasonably fast.

Although physically-based, irradiance caching does not provide an unbiased solution. This is rather a theoretical than a practical problem since the solution is usually indistinguishable from the reference. Another limitation of irradiance caching is its degraded performance in geometrically complex scenes, such as foliage. Indeed, irradiance caching delivers best performance in scenes with not much clutter.

Last but not least, irradiance caching only addresses diffuse to diffuse interreflections. It is of no use for computing global illumination effects involving specular surfaces.

CHAPTER OVERVIEW

The first chapter provides a general introduction to rendering and physically-based lighting simulation. Basic principles of global illumination computation are presented together with their implementation using ray tracing.

Chapter 2 introduces the core of irradiance caching. First, the computation of irradiance and irradiance gradients is detailed. Then, we describe the actual irradiance caching algorithm for reusing computed irradiance values. Several formulas and heuristics are provided to improve the rendering quality. The chapter also discusses the data structure, based on an octree, used to store and fetch irradiance values.

While the irradiance caching algorithm is highly efficient in general, pathological cases may happen frequently especially in the context of production rendering. Chapter 3 provides solutions to such common practical issues. These include the scheme for populating the irradiance cache, as well as the handling of geometric complexity due to bump maps, hair or grass. The chapter also addresses the combination of motion blur with irradiance caching, and the use of irradiance caching for fast ambient occlusion calculation.

As the irradiance caching algorithm does not consider all possible interactions between light and matter, Chapter 4 presents how this algorithm can be integrated within a full global illumination system featuring support for nondiffuse surfaces. The combination with Monte Carlo path tracing and photon mapping will particularly be addressed.

Current graphics processors (GPUs) are generally more powerful than classical processors, as long as their computational model is respected. As the original irradiance caching algorithm cannot be easily implemented on graphics processors, Chapter 5 presents a reformulation of the algorithm meeting the constraints of the GPU. This method especially demonstrates high performance for the computation of one-bounce global illumination.

Generally, the rendering of an animation segment considers all the frames of the animation separately, and performs a separate lighting simulation for each of them. To avoid wasting computation resources, Chapter 6 introduces a solution for reducing the cost of animation rendering by leveraging the temporal coherence of indirect lighting.

ACKNOWLEDGEMENTS

Many thanks to Greg Ward for coming up with irradiance caching; Kadi Bouatouch for encouraging us to write this book; Okan Arikan, Per H. Christensen, Henrik Wann Jensen, Eric Tabellion, and Greg Ward for contributing to the 2007 and 2008 SIGGRAPH courses on irradiance caching; Wojciech Jarosz for many helpful discussions and for correcting mistakes in our papers. Thanks to Kavita Bala, Marko Dabrovic, Henrik Wann Jensen, Wojciech Matusik, Addy Ngan, Greg Ward, Universal Production Partners for giving us the permission to reuse their figures and scenes.

Jaroslav would like to thank to Kavita Bala giving him all the time needed to finish the book during his Cornell stay. Jaroslav acknowledges the support from the Marie Curie grant number PIOF-

GA-2008-221716 and from the Ministry of Education, Youth and Sports of the Czech Republic under the research program LC-06008 (Center for Computer Graphics).

Jaroslav Křivánek
Cornell University and Charles University, Prague

Pascal Gautron
Thomson Corporate Research, France

March 2009

CHAPTER 1

Introduction to Ray Tracing and Global Illumination

Our visual perception of reality is due to the light entering our eyes. But before the light reaches the eyes on its way from the light source, it is usually reflected or refracted many times off object surfaces. In computer graphics, we refer to the light bouncing around in a scene as *global illumination*. Depending on the objects' reflective material characteristics, global illumination creates a wide range of visual phenomena, some of which are shown in Figure 1.1.

In this book, we focus on the computation of one of these phenomena, referred to as *diffuse interreflections* (i.e. light that is reflected diffusely several times before reaching the eyes). Diffuse interreflections create an effect known as *color bleeding*, where one diffuse object "borrows" its color from another. Diffuse interreflections are also responsible for most illumination in interiors (walls reflect light diffusely) and for smooth illumination gradations. The very purpose of the irradiance caching algorithm, described in this book, is to make the computation of diffuse interreflections fast.

This book deals with image rendering, a process that takes a scene description (geometry, materials, light sources, camera) and turns it into an image. We may want to render photo-realistic images, i.e. images that look like a photograph of reality, or create a stylized look for a particular artistic or technical purpose. At any rate, to achieve a certain level of plausibility, it is beneficial to approach the rendering problem as *lighting simulation*, i.e. calculation of light traveling in a scene. This approach is often referred to as *physically-based rendering*.

This chapter presents general concepts of rendering and lighting simulation for the purpose of the book, without paying much attention to rigorousness. More theoretical details are provided in the books by Glassner [Gla95] and Dutré et al. [DBB06]. Practical issues of an implementation of a physically-based ray tracer can be found in the book [PH04]. A general introduction to ray tracing is given in [Gla89]. Mathematical tools and formulas for lighting simulation are summarized in [Dut03]. Appendix A gives a concise overview of spherical geometry, probability, and Monte Carlo integration.

1.1 BASIC RADIOMETRIC QUANTITIES

Radiometry is the field that studies the measurement of electromagnetic radiation, including visible light. See [CW93] for a computer graphics oriented introduction to radiometry. In the context of this book, three basic quantities are discussed: the flux, radiance, and irradiance.

Flux   The *radiant flux* Φ, measured in Watts $[W]$, is the total amount of radiant energy (imagine number of photons) passing through a surface of interest per unit time:

$$\Phi = \frac{dQ}{dt}.$$

(a) refractions

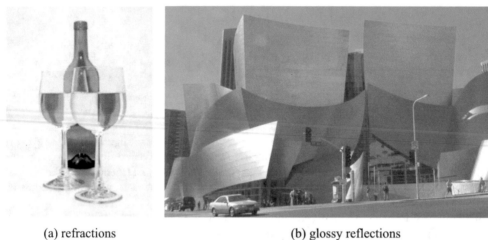

(b) glossy reflections

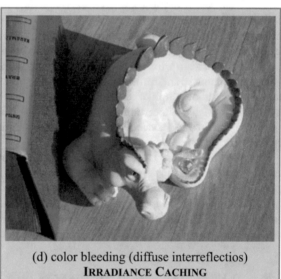

(c) caustics

(d) color bleeding (diffuse interreflectios)
IRRADIANCE CACHING

Figure 1.1: Global illumination causes different visual phenomena depending on the objects's reflectance properties. In this book, we focus on the computation of *diffuse interreflections*, where light reflects diffusely multiple times before reaching the eye.

Irradiance   The *irradiance* E, expressed in W.m^{-2}, is the *incident* flux per unit area:

$$E(\mathbf{p}) = \frac{\mathrm{d}\Phi(\mathbf{p})}{\mathrm{d}A}.$$

Irradiance describes the spatial density of radiation incident at a surface, regardless of the direction from which the radiation arrives. The value of irradiance depends on the orientation of the surface element dA. In particular, only the light incident from the upper hemisphere H^+ around the surface normal at $\mathbf{p}$ contributes to irradiance at $\mathbf{p}$.

Radiance   The *radiance L* localizes flux in space and direction. It is the flux per unit projected area, per unit solid angle:

$$L(\mathbf{p}, \omega) = \frac{\mathrm{d}^2\Phi(\mathbf{p}, \omega)}{\mathrm{d}A\,\mathrm{d}\omega\,\cos\theta} = \frac{\mathrm{d}E(\mathbf{p}, \omega)}{\mathrm{d}\omega\,\cos\theta},$$

where dω is a differential solid angle, and θ is the angle between the surface normal and the direction ω. The cosine term $\cos\theta$ makes the value of radiance independent of the mutual orientation of the surface element dA and the direction of flux ω, since it projects dA into ω.

Irradiance can then be expressed in terms of radiance as:

$$E(\mathbf{p}) = \int_{H^+} L(\mathbf{p}, \omega) \cos\theta\,\mathrm{d}\omega, \tag{1.1}$$

where H^+ is the upper hemisphere centered on the surface normal at $\mathbf{p}$.

1.2   RENDERING GOAL: FIND RADIANCE

Our goal in image rendering is to compute the color of each pixel in the image. In terms of lighting simulation, we want to calculate how much light is reflected from the objects visible through pixels towards the virtual camera.

We use the radiometric quantity *radiance* $L(\mathbf{p}, \omega)$ to measure the "amount of light" reaching or leaving point $\mathbf{p}$ in direction ω. Two fundamental properties make radiance the most important quantity in lighting simulation:

- *The response of the eye or a camera film is proportional to radiance.* This is why our goal in rendering is to compute radiance.

- *Radiance is constant along straight lines in space.* This is why we express the amount of light carried by a ray in terms of radiance.

We distinguish the *outgoing radiance* $L_o(\mathbf{p}, \omega)$, describing light leaving the point $\mathbf{p}$ along ω, and the *incoming radiance* $L_i(\mathbf{p}, \omega)$, which refers to light incident at $\mathbf{p}$.

To summarize, our goal in physically-based image rendering is to determine the outgoing radiance $L_o(\mathbf{p}, \omega_o)$ for every surface point $\mathbf{p}$ visible through image pixels, in the *outgoing direction* ω_o pointing from $\mathbf{p}$ towards the virtual camera (see Figure 1.2).

The outgoing radiance is the sum of the radiance emitted by the surface that $\mathbf{p}$ lies on (in case $\mathbf{p}$ is on an area light source), L_e, and the radiance due to light reflection, L_r:

$$L_o(\mathbf{p}, \omega_o) = L_e(\mathbf{p}, \omega_o) + L_r(\mathbf{p}, \omega_o).$$

The self emission $L_e(\mathbf{p}, \omega_o)$ is easily determined, since it is given in the scene database. In what follows, we focus on the reflected radiance $L_r(\mathbf{p}, \omega_o)$.

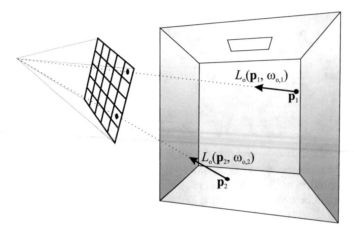

Figure 1.2: Our goal in physically-based image rendering is to determine the outgoing radiance $L_o(\mathbf{p}, \omega_o)$ for every surface point $\mathbf{p}$ visible through image pixels, in the outgoing direction ω_o pointing from $\mathbf{p}$ towards the virtual camera.

1.3   RAY TRACING

Ray tracing [Gla89, PH04] is a particularly simple algorithm for image rendering. For each pixel, we set up a ray[1] that originates at the center of projection and passes through the pixel. The nearest of the intersections of this ray with objects in the scene is the point visible through that pixel. For each of the points, the Shade() procedure, discussed later, computes the color to be assigned to the pixel. The image rendering procedure using ray tracing is summarized in Algorithm 1.

In a naïve approach, the nearest intersection is found by testing the intersection of the ray with each object and choosing the one closest to the ray origin (Figure 1.3(a)). To accelerate the intersection procedure, a spatial data structure, such as a bounding volume hierarchy (BVH) or a kd-tree, is used to quickly cull away objects that cannot be intersected by a given ray (Figure 1.3(b)).

Ray tracing is of particular importance for us since it is at the heart of irradiance caching. However, in film production rendering, it is more common to use rendering algorithms based on the REYES architecture [CCC87], which is for example the basis of PrMan, Pixar's implementation of the RenderMan standard [Ups90]. In REYES, the geometry of the scene is tesselated into small quadrilaterals, called *micropolygons*, the projected size of which is smaller than a pixel. Each micropolygon is then shaded, and the visibility is solved using a variant of the z-buffer algorithm. z-buffer is also used to solve the visibility on the GPU. Nevertheless, no matter what particular rendering algorithm is used, there is always a Shade() procedure that calculates the color (i.e. outgoing radiance) for a point on an object surface.

1.4   SHADING, REFLECTANCE, AND THE BRDF

The shading calculation is what determines the object appearance in the rendered image. For improved flexibility, the shading in many rendering systems is programmable through so called *shaders* (for example RSL shaders in RenderMan or GPU vertex/pixel shaders). A shader simply receives all the information

---

[1]A *ray* is a half line defined by its *origin* and *direction*.

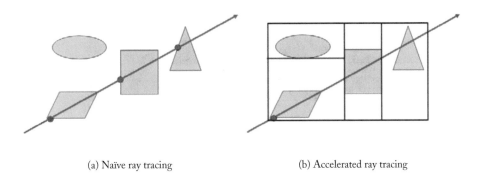

(a) Naïve ray tracing   (b) Accelerated ray tracing

Figure 1.3: Naïve ray tracing performs ray/geometry intersections with all the objects of the scene (a). Using an acceleration data structure based on space partitioning, the nearest intersection is obtained far more efficiently by culling objects that cannot be intersected by a given ray.

about the point being shaded (position, normal, viewing direction, material properties, etc...) and computes the resulting "color", or, in the terminology of physically-based rendering, the outgoing radiance L_o.

As shown in Figure 1.4, the appearance of an object depends on its reflectance characteristics. In computer graphics, we use *shading models* to describe reflectance. Some of the common reflectance models include Lambertian (for purely diffuse objects), Phong or Blinn (add specular highlights). A

Algorithm 1 Image rendering with ray tracing.

---

procedure RayTraceImage
    for all pixels (i, j) in the image **do**
        dir ← direction from camera position to pixel (i, j)
        ray ← [cameraPos, dir]   ▷ *Ray from camera through pixel (i, j)*
        image$[i, j]$ ← Trace(ray)
    end for
end procedure

function Trace(ray)   ▷ *Determine radiance for a given ray*
    hitInfo ← Intersect(ray)   ▷ *Find the nearest intersection*
    if hitInfo.object $\neq$ NULL **then**
        return Shade(hitInfo, ray)   ▷ *Calculate radiance*
    else
        return Background(ray)   ▷ *Use background if no hit*
    end if
end function

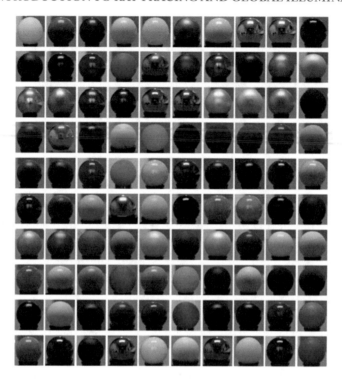

Figure 1.4: Appearance of objects depends on their reflectance characteristics. All the spheres are illuminated by the same lighting environment, only their material reflectance differs. (Image courtesy of Wojciech Matusik.)

shading model is a mathematical abstraction that describes an object's response to incoming light. In practice, it is expressed in terms of a formula that calculates light intensity at a point.

In physically-based rendering, the equivalent of the shading model is the *bi-directional reflectance distribution function* or the BRDF, usually denoted f_r. At a given point, the BRDF is a function of two directions, the incoming ω_i and the outgoing ω_o (Figure 1.5). It is defined as the ratio between the reflected outgoing radiance L_r and the differential irradiance E:

$$f_r(\omega_i, \omega_o) = \frac{dL_r(\omega_o)}{dE(\omega_i)} = \frac{dL_r(\omega_o)}{L_i(\omega_i)\cos\theta_i\,d\omega_i}, \tag{1.2}$$

where L_i is the incoming radiance and θ_i is the angle between the incoming light direction ω_i and the surface normal at the point where the BRDF is evaluated. Think of the BRDF as the percentage of the light coming from ω_i that gets reflected towards ω_o.

The cosine term $\cos\theta_i$ deserves a special attention since it keeps appearing in most of our formulas. Imagine a flashlight shining on a surface. As we rotate the flashlight away from the surface normal, the brightness (which is our perception of the reflected radiance) of a fixed point on the surface decreases, even though the incoming radiance remains the same (see Figure 1.6). This is quite logical, since the

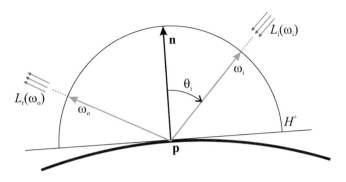

Figure 1.5: The BRDF at a point **p** gives the percentage of the radiance coming from the incoming direction ω_i that is reflected towards the outgoing direction ω_o.

photons emitted by the flashlight spread over a larger area, so their density at any given point will be lower. The decrease in reflected radiance is inversely proportional to the area illuminated by the flashlight, i.e. exactly $\cos\theta_i$. This decrease of reflected radiance is exclusively due to the geometric configuration and has nothing to do with the reflectance properties of the surface. This is why the cosine term $\cos\theta_i$ was introduced in the definition of the BRDF to factor out this purely geometric effect.

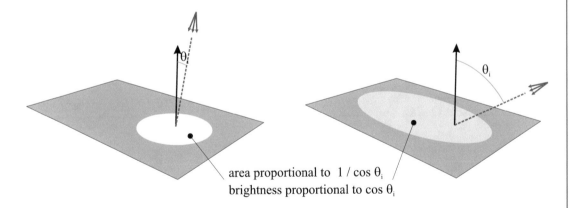

Figure 1.6: Decrease in brightness (i.e. the reflected radiance) is proportional to the cosine term $\cos\theta_i$, irrespective of the surface reflectance properties. That is why the cosine term is factored out in the definition of the BRDF.

Many BRDF models have been devised to describe different material types. Among those, common BRDFs include Lambert model for diffuse surfaces, Phong/Blinn [Pho75] for glossy reflections, or Ward [War92] for anisotropic materials. Fixing the incident direction ω_i, the BRDF is a function of only the outgoing direction ω_o (and vice-versa). We often refer to the BRDF for a fixed ω_i or ω_o as the BRDF *lobe*. Figure 1.7 shows a plot of BRDF lobes for the aforementioned BRDFs.

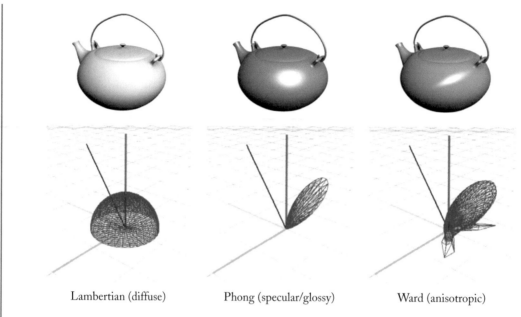

Lambertian (diffuse)   Phong (specular/glossy)   Ward (anisotropic)

Figure 1.7: Renderings and plots of common BRDF models.

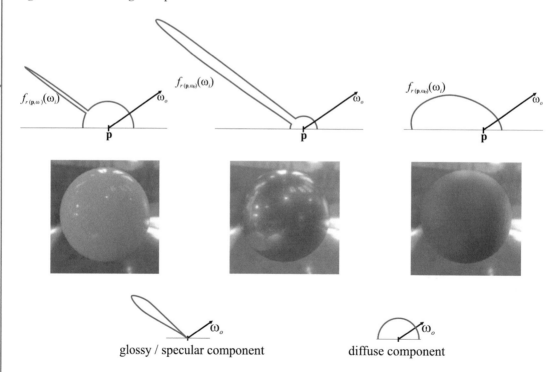

glossy / specular component   diffuse component

Figure 1.8: Many real-world BRDF have distinct diffuse and specular/glossy components. For some BRDFs, on the other hand, the ideal diffuse component may not be present at all. (Images courtesy Addy Ngan.)

Most of the materials in the real world tend to reflect light in specific directions, yielding *view-dependent effects*. That is to say, what we see on an object depends on where we look at it from. Such surfaces are generally characterized as more or less *specular* or *glossy* depending on this tendency.

A special case of the view-dependent reflection is *ideal (perfect) specular reflection*. Given the incident direction, there is only a single outgoing direction that the light is reflected to. In other words, the BRDF lobe is zero except for one single direction. This is how a mirror reflects light.

When the surface tends to reflect the light uniformly in every direction, this surface is called *diffuse*. For an ideal diffuse surface, also called *Lambertian*, the BRDF is a constant function,

$$f_r(\omega_i, \omega_o) = \frac{\rho_d}{\pi},$$

where $\rho_d \in [0, 1]$ is called the *diffuse reflectivity* of the material. This is what we perceive as the surface color.

Many real-world materials feature both the specular/glossy and diffuse reflections, as shown in Figure 1.8. For practical purposes, we usually model such BRDFs as a sum of *components*: diffuse and glossy or ideal specular. This allows us to use different algorithms to solve illumination due to each of the components. For example, irradiance caching only applies to the diffuse component of a BRDF.

1.5   DIRECT AND INDIRECT ILLUMINATION

Now we know how light is reflected when it arrives at a surface but we still have not discussed where the light actually comes from. First, *direct illumination* is due to the light that comes directly from the light sources (Figure 1.9(a,b)). However, light arrives also indirectly, after multiple reflections on other scene surfaces; this light produces *indirect illumination* (Figure 1.9(c,d)). The sum of direct and indirect illumination is *global illumination*:

$$\text{global illumination} = \text{direct illumination} + \text{indirect illumination}.$$

In this book, we focus on the computation of *indirect illumination on diffuse surfaces*.

1.5.1   DIRECT ILLUMINATION

In the case of direct illumination we know that the light comes from the light sources and we know exactly where they are located and how much light they emit (this information is given in the scene database). This makes direct illumination computation simple. Just iterate over the lights, multiply the incoming radiance due to each light by the BRDF and the cosine term to turn it into the corresponding reflected radiance, and sum it all together:

$$L_r^{\text{direct}}(\mathbf{p}, \omega_o) = \sum_{k=1}^{\#lights} L_{i,k}(\mathbf{p}, \omega_{i,k}) f_r(\mathbf{p}, \omega_{i,k}, \omega_o) \cos \theta_{i,k}$$

In the above formula, we have explicitly used $\mathbf{p}$ as one of the BRDF parameters since in general each point in the scene may have a different BRDF.

Naturally, light sources that are occluded by other scene objects do not contribute to the direct illumination since the occluding objects cast shadows. In ray tracing, the shadow computation involves testing the intersection of a *shadow ray* from $\mathbf{p}$ to the light source with scene objects (Figure 1.9(a)). If an intersection is detected, the light does not contribute to direct illumination. Algorithm 2 gives a pseudocode for direct illumination computation with shadows using ray tracing.

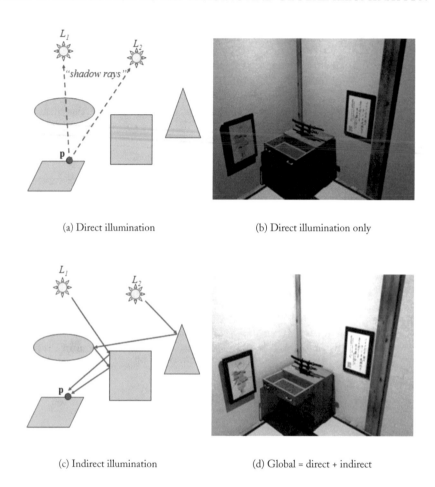

(a) Direct illumination

(b) Direct illumination only

(c) Indirect illumination

(d) Global = direct + indirect

Figure 1.9: Direct illumination is due to the light reaching the observed point **p** directly from the light sources. In ray tracing, it can be computed by looping over the light sources and testing for occlusion using the shadow rays. In (a), light source L_1 is occluded while source L_2 contributes to the illumination at **p**. However, in reality, light reaches **p** also after multiple reflections in the scene, yielding indirect illumination (c).

1.5.2   INDIRECT ILLUMINATION

1.5.2.1   Ideal Specular Reflections

Given the outgoing (or viewing) direction ω_o, there is a single direction ω_i from which the light is reflected by the ideal (perfect) specular reflection. This makes the computation of indirect illumination on an ideal specular surface simple. We cast a *secondary ray* in the direction of specular reflection. The outgoing radiance computed at the point where the secondary ray hits the scene gives us the indirect illumination. To compute this outgoing radiance, ray tracing is applied recursively as shown in Algorithm 3. (The

Direct illumination only   Global = direct + indirect

Figure 1.10: Another example of the difference between the direct and global illumination.

coefficient k_r in the algorithm is a surface property that gives the percentage of light reflected by specular reflection. In reality, k_r is the function of ω_o as described by the Fresnel equations [PH04]. In practice, however, it is often set directly by the user.)

Algorithm 2 Direct illumination in ray tracing.

---

    function DirectIllumination(hitInfo, incidentRay)
        $\omega_o \leftarrow -$incidentRay.direction   ▷ ω_o *points away from the surface*
        $\mathbf{p} \;\leftarrow$ hitInfo.position
        $L_r \leftarrow 0$
        for all light sources k **do**
            $\omega_{i,k} \leftarrow$ direction from $\mathbf{p}$ to light source k
            $L_{r,k} \leftarrow L_{i,k}(\mathbf{p}, \omega_{i,k}) f_r(\mathbf{p}, \omega_{i,k}, \omega_o) \cos \theta_{i,k}$
            if $\|L_{r,k}\| > 0$ **then**   ▷ *shadow test only for non-zero contributions*
                shadowRay $\leftarrow [\mathbf{p}, \omega_{i,k}]$
                maxDist $\leftarrow$ distance from $\mathbf{p}$ to light k
                if not HasIntersection(shadowRay, maxDist$-\varepsilon$) **then**
                    L_r += $L_{r,k}$ ▷ *use contribution if not shadowed*
                end if
            end if
        end for
        return L_r
    end function

Algorithm 3 Specular reflection using recursive ray tracing.

function SpecularReflection(hitInfo, incidentRay)

    $\omega_o \leftarrow -$incidentRay.direction   ▷ *ω_o points away from the surface*

    p $\leftarrow$ hitInfo.position

    $\omega_i \leftarrow$ ReflectedDirection(hitInfo, ω_o)

    secondaryRay $\leftarrow [\mathbf{p}, \omega_i]$

    return $k_r \cdot$ Trace(secondaryRay)

end function

1.5.2.2   General BRDFs: Distribution Ray Tracing

In the case of a general BRDF, the situation is more complicated than for an ideal specular reflection since the light from the whole hemisphere of incoming directions ω_i can be reflected towards an outgoing direction ω_o. However, we can still use tracing of secondary rays to determine indirect illumination. We generate a number of random directions on the hemisphere according to a selected probability distribution and cast secondary rays in these directions. The outgoing radiance computed at the hit point of each secondary ray gives us the incoming radiance from the respective direction. The total reflected radiance is then computed using a simple average:

$$L_r^{\text{indirect}}(\mathbf{p}, \omega_o) = \frac{1}{N} \sum_{k=1}^{N} \frac{L_{i,k}(\mathbf{p}, \omega_{i,k}) f_r(\mathbf{p}, \omega_{i,k}, \omega_o) \cos \theta_{i,k}}{p(\omega_{i,k})}, \tag{1.3}$$

where

N is the number of *samples*, i.e. secondary rays,

$\omega_{i,k}$ is the kth random sample direction, and

$p(\omega_{i,k})$ is the value of the probability density function (PDF) for generating random direction $\omega_{i,k}$.

In short, we estimate the reflected radiance by *sampling* the incoming radiance field. Each of the samples corresponds to casting a secondary ray. Algorithm 4 gives a pseudocode for the indirect illumination computation.

    Since we distribute the secondary rays over the hemisphere, the technique is often referred to as *distribution ray tracing*. Its other name, *stochastic ray tracing*, follows from the stochastic nature of the algorithm: the secondary rays are cast in random directions. Yet another name for the same thing, *Monte Carlo ray tracing*, stems from the fact that we are using Monte Carlo integration, as becomes clear shortly.

    Due to the stochastic nature of distribution ray tracing we can see *noise* in the resulting images. The noise can be reduced by increasing the number of samples. However, it only decreases with the square root of the number of samples: To reduce noise twice, we need four times as many rays (see Figure 1.11).

    A more effective way of reducing the noise is *importance sampling*. Consider computing indirect illumination on a glossy surface. Since there is a strong tendency to reflect light only in some directions, we cast the rays preferably in those directions (see Figure 1.12). In other words, the probability density function (PDF) for generating the random directions is proportional to the BRDF lobe (for the given fixed outgoing direction). Importance sampling can reduce noise without casting extra rays.

    However, importance sampling is not very effective on diffuse surfaces, since the light is reflected almost equally from all directions. Therefore, noise reduction requires increasing the number of secondary

---

Algorithm 4 Indirect illumination using distribution (stochastic) ray tracing.

---

    function IndirectIllumination(hitInfo, incidentRay, N)

        $\omega_o \leftarrow -$incidentRay.direction   ▷ ω_o *points away from the surface*

        $\mathbf{p} \leftarrow$ hitInfo.position

        $L_r \leftarrow 0$

        for $k \leftarrow 1$ **to** N **do**

            $[\omega_{i,k}, \text{pdf}] \leftarrow$ RandomDirection(hitInfo, ω_o)

            secondaryRay $\leftarrow [\mathbf{p}, \omega_{i,k}]$

            $L_{i,k} \leftarrow$ Trace(secondaryRay)

            L_r += $\frac{1}{\text{pdf}} \cdot L_{i,k} \cdot f_r(\mathbf{p}, \omega_{i,k}, \omega_o) \cdot \cos\theta_{i,k}$

        end for

        return $\frac{L_r}{N}$

    end function

---

rays, which makes the computation very slow. (Usually, we need 100 or more secondary rays per pixel for noise-free images.)

The *irradiance caching* algorithm addresses the problem of slow computation of indirect illumination on diffuse surfaces. The main idea is to perform the costly hemisphere sampling only at a selected set of locations in the scene, store the results in a cache, and reuse the cached value at other points through a fast interpolation.

1.6 ILLUMINATION INTEGRAL AND RENDERING EQUATION

Let us now formalize the reflection of light and indirect illumination computation using the radiometric terms.

1.6.1 ILLUMINATION INTEGRAL

Considering the definition of the BRDF (1.2), we can write the differential reflected radiance dL_r as $dL_r(\mathbf{p}, \omega_o) = f_r(\mathbf{p}, \omega_i, \omega_o) L_i(\mathbf{p}, \omega_i) \cos\theta_i \, d\omega_i$. To get the total outgoing radiance, we sum the contributions of the incoming radiance from all directions on the hemisphere, i.e. we integrate over the hemisphere:

$$L_r(\mathbf{p}, \omega_o) = \int_{H^+} L_i(\mathbf{p}, \omega_i) f_r(\mathbf{p}, \omega_i, \omega_o) \cos\theta_i \, d\omega_i. \tag{1.4}$$

The above integral is referred to as the reflection equation or the *illumination integral* (we use the latter term in this book). The illumination integral gives us the total reflected radiance at a point given the incoming radiance from all directions and the BRDF.

Formally, Equation (1.3) is a *Monte Carlo estimator* for numerical evaluation of the illumination integral. See Appendix A for a short general introduction to Monte Carlo integration.

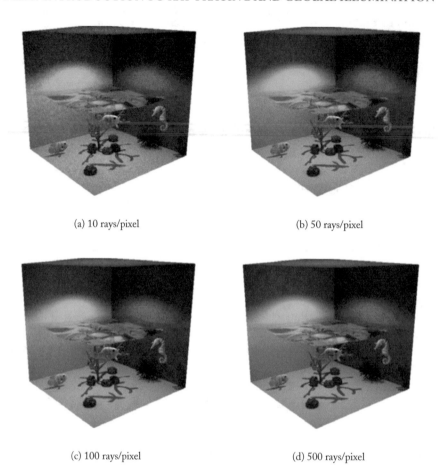

(a) 10 rays/pixel   (b) 50 rays/pixel

(c) 100 rays/pixel   (d) 500 rays/pixel

Figure 1.11: Noise in distribution ray tracing decreases with the square root of the number of samples (i.e. rays).

1.6.2   LAMBERTIAN REFLECTION

The use of the ideal diffuse, or Lambertian, BRDF greatly simplifies the lighting simulation. Recall that the Lambertian BRDF is constant in ω_i and ω_o, i.e. $f_r(\mathbf{p}, \omega_i, \omega_o) = \rho_d(\mathbf{p})/\pi$, where $\rho_d \in [0, 1]$ is the diffuse reflectivity. We can now simplify the illumination integral as follows:

$$L_r(\mathbf{p}, \omega_o) = \frac{\rho_d(\mathbf{p})}{\pi} \int_{H^+} L_i(\mathbf{p}, \omega_i) \cos \theta_i \, d\omega_i$$

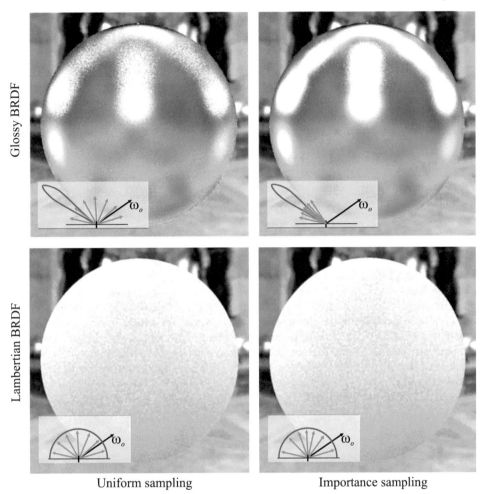

Figure 1.12: Importance sampling strategy reduces image noise by directing secondary rays where the BRDF times the cosine term is large. This techniques is very effective for glossy BRDFs (top row). However, for diffuse BRDFs (bottom row), the noise reduction is low, since light is reflected from all directions (importance sampling for the Lambertian surface only uses the cosine term).

Using the relation of irradiance and incoming radiance $E(\mathbf{p}) = \int_{H^+} L_i(\mathbf{p}, \omega_i) \cos\theta_i \, d\omega_i$, we express the reflected radiance in terms of the irradiance:

$$L_\mathrm{r}(\mathbf{p}, \omega_\mathrm{o}) = \frac{\rho_d(\mathbf{p})}{\pi} E(\mathbf{p})$$

Reflection on an ideal diffuse surface is said to be *view-independent* since the outgoing radiance is the same in all outgoing (or viewing) directions. The view independence of ideal diffuse reflection is a strong asset for irradiance caching: a single color value, the irradiance, determines the amount of reflected light for any viewing direction.

1.6.3   RENDERING EQUATION

The illumination integral expresses the reflected radiance at a point as an integral of incoming radiance and the BRDF over the hemisphere. However, it does not tell us anything about the incoming radiance itself.

Using the fact that radiance is constant along straight lines, we have

$$L_i(\mathbf{p}, \omega_i) = L_o(\text{isect}(\mathbf{p}, \omega_i), -\omega_i),$$

where the function $\text{isect}(\mathbf{p}, \omega_i)$ returns the nearest intersection of a ray from $\mathbf{p}$ in the direction ω_i. Plugging back to the illumination integral yields the *Rendering Equation* [Kaj86]:

$$L_o(\mathbf{p}, \omega_o) = L_e(\mathbf{p}, \omega_o) + \int_{H^+} L_o(\text{isect}(\mathbf{p}, \omega_i), -\omega_i) f_r(\mathbf{p}, \omega_i, \omega_o) \cos\theta_i \, d\omega_i. \qquad (1.5)$$

Note that we have reintroduced the self-emission $L_e(\mathbf{p}, \omega_o)$. Although similar to the illumination integral in its form, the Rendering Equation has a very different meaning. It expresses radiance at a point and direction in the scene in terms of radiance at different points and directions in the scene. The unknown, radiance L_o, appears on both sides of the equation (which makes it a true integral equation, whereas the illumination integral is simply a formula for computing the amount of light reflected at a point).

A number of techniques exist to solve the Rendering Equation. The distribution ray tracing adopted in this book solves the rendering equation as an infinite series of recursive evaluations of the illumination integral (the recursion is, of course, limited in practice).

CHAPTER 2

Irradiance Caching Core

As described in the previous chapter, a straightforward way to compute indirect illumination at a point on a diffuse surface is hemisphere sampling: a number of secondary rays are traced and their radiance contributions are averaged. However, doing so at each visible point is prohibitive, since many rays must be traced to obtain results free of visible noise.

Irradiance caching decreases the overall cost of indirect illumination computation by performing hemisphere sampling only at selected points in the scene, caching the results, and reusing the cached indirect illumination values between these points through interpolation. The algorithm can be summarized as follows:

> **if** interpolation is possible **then**
>   reuse cached values through interpolation
> **else**
>   compute new value and store it in the cache
> **end if**

Figure 2.1: The left image shows a global illumination rendering of a conference hall. The indirect (RGB) irradiance, shown on the right, tends to change slowly overall, especially on flat surfaces in open spaces. It changes more rapidly on curved surfaces and in the vicinity of scene geometry. (Images courtesy of Greg Ward.)

This approach is made possible by the spatial coherence of indirect illumination on diffuse surfaces: indirect illumination changes quite slowly over surfaces, as illustrated in Figure 2.1.

The indirect illumination "value" stored in the cache is the *irradiance*, $E(\mathbf{p})$. Since irradiance describes the total amount of light incident at a point irrespective of its directionality, irradiance caching is limited to view-independent, purely diffuse (Lambertian) reflection.

The rest of this chapter gives all the details necessary to turn the above simplified procedure into an actual implementation. In Section 2.1, we start by a detailed description of irradiance calculation at a point. The same section also covers the computation of *irradiance gradients* since they are indeed computed together with a new irradiance value. The gradients substantially improve image quality when used in interpolation. Once computed, the irradiance and its gradients, along with other information, such as position, normal, etc. are stored in the cache as a new *record*.

The true core of irradiance caching is the determination of the records that can be reused at a given point and the interpolation procedure itself. These issues, discussed in Section 2.2, also imply the details on the actions taken when constructing a new cache record, as described in the same section.

Fast storage and retrieval of the cache records is a vital part of irradiance caching. The data structures and algorithms used for this purpose are discussed in Section 2.3. Finally, Section 2.4 summarizes the algorithm and gives references to particular sub-algorithms and formulas introduced throughout the chapter.

2.1   INDIRECT IRRADIANCE CALCULATION

Irradiance at a point $\mathbf{p}$ with normal $\mathbf{n}$ is given by the integral of incoming radiance over the hemisphere:

$$E(\mathbf{p}) = \int_{H^+} L_i(\mathbf{p}, \omega) \cos\theta \, d\omega \qquad (2.1)$$

where H^+ is the upper hemisphere above $\mathbf{p}$, $L_i(\mathbf{p}, \omega)$ is the incoming radiance reaching $\mathbf{p}$ from direction ω. The spherical coordinates of ω are denoted (θ, ϕ).

Instead of trying to compute the value of this integral exactly, which is anyway impossible in most practical cases, we estimate it numerically using Monte Carlo stratified importance sampling. *Stratification* subdivides the hemisphere into cells and chooses one random direction in each cell (see Figure 2.2). Secondary rays are traced in these directions, and a shading calculation at the hit points of these rays returns the incoming radiance samples from which the irradiance is estimated by averaging. From the point of view of irradiance caching, it does not matter much how exactly the incoming radiance samples are calculated, so we will postpone the discussion of radiance samples calculation to Section 4.3. One important thing to keep in mind, though: Since we are estimating *indirect* illumination, any light directly emitted by the intersected object (in case we hit a light source) is not taken into account.

Importance sampling is used to slightly reduce variance of our irradiance estimate. Recall from Section 1.5.2.2 that importance sampling places more samples where the integrand is large. Since nothing is à priori known about the incoming radiance in our case, only the cosine term $\cos\theta$ from the irradiance integral (2.1) is used as the importance function, i.e. samples are more likely to be placed around the surface normal (where $\cos\theta = 1$) than near the tangent plane (where $\cos\theta = 0$). The cosine-proportional sampling strategy is equivalent to distributing samples uniformly on the projection of the hemisphere on the tangent plane. This is why the cells due to stratification all have the same area in the projection (see Figure 2.2).

Formally, we draw the samples from a distribution with the probability density function (PDF) given by:

$$p(\theta, \phi) = \frac{\cos\theta}{\pi}.$$

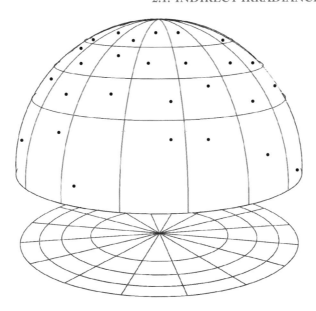

Figure 2.2: Hemisphere sampling for indirect irradiance estimation. Stratification divides the hemisphere into cells and generates one sample in each cell. The cells projected to the tangent plane all have the same area.

The division by π makes the function integrate to 1 on the hemisphere, which is a general requirement on any probability density. The general form of the stratified Monte Carlo estimator is:

$$E(\mathbf{p}) \approx \frac{1}{MN} \sum_{j=0}^{M-1} \sum_{k=0}^{N-1} \frac{f(\theta_{j,k}, \phi_{j,k})}{p(\theta_{j,k}, \phi_{j,k})},$$

where f is the integrand and p the PDF for drawing samples. In our case $f(\theta, \phi) = L(\theta, \phi) \cos \theta$ and $p(\theta, \phi) = \frac{\cos \theta}{\pi}$, which yields the final *irradiance estimator*:

$$E(\mathbf{p}) \approx \frac{\pi}{MN} \sum_{j=0}^{M-1} \sum_{k=0}^{N-1} L_{j,k}, \tag{2.2}$$

where

$L_{j,k}$ is the incoming radiance sample computed by tracing a ray in the direction

$$(\theta_{j,k}, \phi_{j,k}) = \left(\arccos \sqrt{1 - \frac{j + \zeta_{j,k}^{1}}{M}}, \; 2\pi \frac{k + \zeta_{j,k}^{2}}{N} \right). \tag{2.3}$$

In cartesian coordinates

$$x_{j,k} \;=\; \sin\theta_{j,k}\cos\phi_{j,k} \;=\; \sqrt{\frac{j+\zeta^1_{j,k}}{M}}\,\cos 2\pi\,\frac{k+\zeta^2_{j,k}}{N}$$

$$y_{j,k} \;=\; \sin\theta_{j,k}\sin\phi_{j,k} \;=\; \sqrt{\frac{j+\zeta^1_{j,k}}{M}}\,\sin 2\pi\,\frac{k+\zeta^2_{j,k}}{N}$$

$$z_{j,k} \;=\; \cos\theta_{j,k} \;=\; \sqrt{1-\frac{j+\zeta^1_{j,k}}{M}}.$$

(The above formulas distribute the directions proportionally to the cosine term [PH04].)

$\zeta^1_{j,k}, \zeta^2_{j,k}$ are two uniformly distributed random numbers in the range [0, 1),

$N \cdot M$ is the total number of sample directions (usually several hundred or thousand). The number of hemisphere divisions along θ is denoted M, and N denotes the number of divisions along ϕ. To obtain similar division density along θ and ϕ, it is advisable to choose M and N such that $N \approx \pi M$.

Weighting of incoming radiance by the cosine term present in the irradiance integral (2.1) is not made explicit in the estimator (2.2) since the importance sampling strategy implicitly takes care of the cosine weighting.

 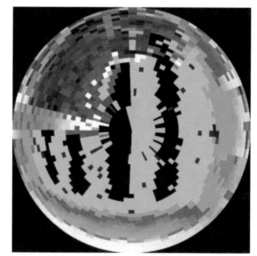

(a) hemispherical fisheye view   (b) incoming radiance samples

Figure 2.3: (a) A hemispherical fisheye view of a conference room scene from a point on the floor. (b) Incoming radiance samples generated by our stratified sampling strategy. The light sources appear dark in (b) since direct light emission from the intersected objects is ignored in indirect illumination sampling. (Images courtesy of Greg Ward.)

Figure 2.3 (a) shows a conference room scene as seen through a hemispherical fisheye lens from a point on the floor. Figure 2.3 (b) shows the values for 2000 incoming radiance samples generated by the stratified sampling procedure described above. The light sources on the ceiling appear dark since light emitted directly by the intersected objects is ignored in indirect illumination calculation.

Stratification avoids clumping of many sample directions in small areas which usually reduces variance of the irradiance estimate compared to purely random sampling without stratification. However, other sampling strategies, such as quasi-Monte Carlo sampling, may provide more effective variance reduction [Kel03, MFS04]. We stick with simple stratification because it proves to be useful for the gradient estimation described in the next section.

Illumination-Aware Hemisphere Sampling   The hemisphere sampling procedure as described so far is simple but unaware of the illumination in the scene. Ideally, we would like to send more rays in directions corresponding to the bright parts of the scene or areas with high illumination variance. For that purpose, Jensen [Jen01] suggests to use *importance sampling* based on the information stored in a photon map. The implementation of *Radiance* uses *adaptive sampling*: an initial batch of samples is shot, variance of incident radiance is determined for these samples, and additional samples are shot only in the directions with high variance [WLS98].

2.1.1   IRRADIANCE GRADIENTS

Hemisphere sampling gives us an estimate of irradiance at a point, which is stored in the cache to be later reused at different locations with different surface normals. If we could predict how the computed irradiance value changes as we move over a surface or change its orientation, we could obtain more accurate interpolation results. Ward and Heckbert [WH92] have shown that a first-order approximation of the irradiance change, i.e. irradiance gradients, can be estimated directly from the hemisphere samples with only a negligible additional cost.

Irradiance in a scene is a five-dimensional scalar field (three dimensions for position and two for surface orientation). In general, gradient of such a field is a five-dimensional vector. For computational convenience we represent this five-dimensional gradient as two independent three-dimensional vectors: the rotation and the translation gradients. The two gradients lie in the base plane of hemisphere sampling, i.e. the tangent plane of the surface, so each of them in fact only represents two degrees of freedom. Note that there is one gradient vector for each color component.

2.1.1.1   Rotation Gradient

The rotation gradient tells us how irradiance changes with rotation. Consider the example in Figure 2.4 (a). A bright surface in the scene contributes indirect light to a surface element. As we rotate the surface element so that its normal points towards the bright surface, irradiance at the element will increase because the contribution from the bright surface will get promoted by the cosine weighting.

The rotation gradient gives us a first order approximation of the irradiance change with rotation (see Figure 2.4 (b)). Its *direction* is the axis of rotation that induces the fastest change in irradiance. Its *magnitude* expresses how quickly the irradiance changes: it is the irradiance derivative with respect to rotation around the axis given by the gradient direction.

The rotation gradient is estimated simultaneously with new irradiance value computation, from the same set of radiance samples $L_{j,k}$, using the following formula:

$$\nabla_r E \approx \frac{\pi}{MN} \sum_{k=0}^{N-1} \left(\mathbf{v}_k \sum_{j=0}^{M-1} -\tan\theta_j \cdot L_{j,k} \right), \tag{2.4}$$

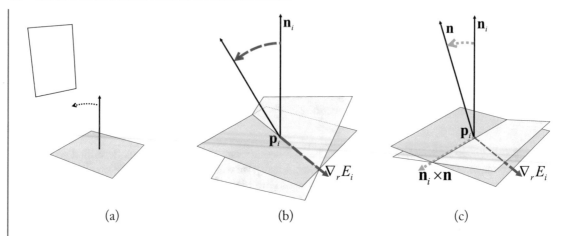

(a)   (b)   (c)

Figure 2.4: (a) As the surface element is rotated towards the bright surface, irradiance increases. (b) The rotation gradient $\nabla_r E_i$ of cache record i gives the axis of rotation that produces maximum increase in irradiance. The gradient magnitude is the irradiance derivative with rotation around that axis. (c) When the surface element is rotated around any arbitrary axis (in our example determined by the change in surface normal as $\mathbf{n}_i \times \mathbf{n}$) the irradiance derivative is given by the dot product of the axis of rotation and the rotation gradient: $(\mathbf{n}_i \times \mathbf{n}) \cdot \nabla_r E_i$.

where

$\mathbf{v}_k$ is a base-plane vector in the direction $\phi_k + \frac{\pi}{2}$ (see Figure 2.5),
(θ_j, ϕ_k) are the spherical coordinates of the center of the hemisphere cell (j, k),
$L_{j,k}$ is the incoming radiance sample computed by tracing a ray through cell (j, k).

A derivation of the rotation gradient formula is given in Appendix B.

Having the gradient vector, we can now approximate how irradiance changes with a rotation of the surface around an arbitrary axis. A change of surface normal, shown in Figure 2.4 (c), is equivalent to a rotation of the surface element around the axis given by the cross product of the normal of record i and the normal at the interpolation location $(\mathbf{n}_i \times \mathbf{n})$. The first-order approximation of the change of irradiance with this rotation is given by the dot product of the actual rotation axis and the rotation gradient vector of the record: $(\mathbf{n}_i \times \mathbf{n}) \cdot \nabla_r E_i$. In irradiance caching, this formula is used to extrapolate an irradiance value when reused at a point with a different surface normal (Equation 2.7).

2.1.1.2 Translation Gradient

Translation gradient tells us how irradiance changes as we move over a surface away from the point where it was calculated by hemisphere sampling (Figure 2.6 (a)). Unlike for rotation gradient, the parallax effect makes the rate of change dependent on the distance to the surfaces that contribute indirect illumination. As we move over a surface, the projection of the surrounding environment onto the hemisphere moves as well. For distant surfaces, the movement is relatively smaller than for near ones. In addition, occlusion and disocclusion play an important role for the translation gradient. For these reasons, we utilize the information about the ray lengths in hemisphere sampling for gradient estimation.

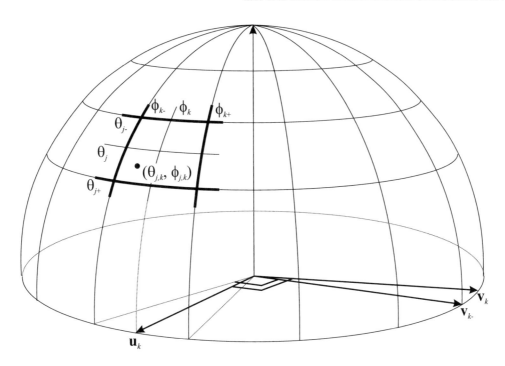

Figure 2.5: Geometry used for gradient estimation.

Note that our translation gradient always lies in the tangent plane of the surface. In other words, we ignore the change of irradiance with translation along the surface normal. This makes perfect sense, since we only interpolate irradiance over surfaces.

The translation gradient is given by the following formula:

$$
\nabla_t E \approx \sum_{k=0}^{N-1} \left[\; \mathbf{u}_k \frac{2\pi}{N} \sum_{j=1}^{M-1} \frac{\cos^2 \theta_{j-} \sin \theta_{j-}}{\min\{r_{j,k}, r_{j-1,k}\}} (L_{j,k} - L_{j-1,k}) + \right.
$$

$$
\left. \mathbf{v}_{k-} \sum_{j=0}^{M-1} \frac{\cos \theta_j (\cos \theta_{j-} - \cos \theta_{j+})}{\sin \theta_{j,k} \min\{r_{j,k}, r_{j,k-1}\}} (L_{j,k} - L_{j,k-1}) \right].
$$

(2.5)

The notation is illustrated in Figure 2.5 and summarized below. The gradient formula is derived in Appendix B.

(j, k) is the cell index,
$L_{j,k}$ is the incoming radiance sample computed by tracing a ray through cell (j, k),
$L_{j,k}$ is the intersection distance of the ray traced through cell (j, k),
θ_{j-} is the elevation angle at the boundary between the current cell (j, k) and the previous cell $(j - 1, k)$,

$$
\theta_{j-} = \arccos \sqrt{1 - \tfrac{j}{M}},
$$

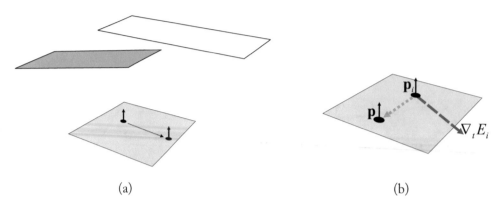

(a)   (b)

Figure 2.6: (a) As the surface element is translated, it becomes more exposed to the bright surface, and irradiance increases. (b) The translation gradient $\nabla_t E_i$ of record i gives the direction of translation that produces the maximum increase in irradiance. The gradient magnitude is the irradiance derivative with respect to translation along that direction. When a surface element is translated along any arbitrary direction, a first-order approximation of the change in irradiance is given by the dot product of the translation vector and the translation gradient: $(\mathbf{p} - \mathbf{p}_i) \cdot \nabla_t E_i$.

θ_{j_+} is the elevation angle at the boundary between the current cell (j, k) and the next cell $(j + 1, k)$,

$$\theta_{j_+} = \arccos \sqrt{1 - \frac{j+1}{M}},$$

ϕ_{k_-} is the azimuthal angle at the boundary between the current cell (j, k) and the previous cell $(j, k - 1)$,

$$\phi_{k_-} = 2\pi \frac{k}{N},$$

ϕ_k is the azimuthal angle at the center of the current cell (j, k), $\phi_k = 2\pi \frac{k+0.5}{N}$,

ϕ_{k_+} is the azimuthal angle at the boundary between the current cell (j, k) and the next cell $(j, k + 1)$,

$$\phi_{k_+} = 2\pi \frac{k+1}{N},$$

$\mathbf{u}_k$ is the base plane unit vector in direction $(\pi/2, \phi_k)$,

$\mathbf{v}_{k_-}$ is the base plane unit vector in direction $(\pi/2, \phi_{k_-} + \pi/2)$.

The translation gradient gives us the direction of the fastest change of irradiance. Its magnitude is the derivative of irradiance with respect to this direction (see Figure 2.6 (b)). Given a displacement along an arbitrary vector, we can approximate the actual change in irradiance as a dot product of the displacement vector and the translation gradient. In irradiance caching, this formula is used to extrapolate irradiance when reused at a different location (Equation (2.7)).

Note that the gradient formulas (2.4) and (2.5) give the gradients in the local coordinate frame at the record location. To avoid frame transformations during irradiance interpolation, we transform the gradients into the global coordinate frame before storing them in the cache.

2.2   IRRADIANCE CACHING ALGORITHM

Irradiance caching accelerates the computation of indirect illumination by performing the hemisphere sampling only at a sparse set of locations in the scene. The computed irradiance values are stored in a

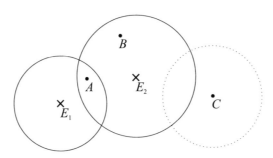

Figure 2.7: Example of lazy irradiance evaluation in irradiance caching. Suppose the irradiance values E_1 and E_2 were computed and stored in the cache previously. The weighted average of E_1 and E_2 is used at a query point A. Query point B uses E_2. Point C is outside the validity radius of both E_1 and E_2, so a new irradiance value is computed and stored as a new record it in the cache. (Figure redrawn after [WLS98].)

cache and reused for fast interpolation at other locations. This section describes in detail when and how the cached values are used for interpolation and when new values are computed and stored in the cache.

The overall caching scheme is based on "lazy" evaluation of irradiance as described in Algorithm 5. New irradiance values are computed on the fly only if none of the cached values can be used for interpolation. An example situation for irradiance caching is shown in Figure 2.7.

The IrradianceCaching() function from Algorithm 5 is called whenever we want to calculate diffuse indirect illumination, e.g. when a primary ray in a ray tracer hits a surface. The returned irradiance value is then simply multiplied by the diffuse reflectance (e.g. surface color or texture value) to get the outgoing radiance contribution due to diffuse indirect illumination:

$$L_o^{\text{diff\_ind}}(\mathbf{p}) = \frac{\rho_d(\mathbf{p})}{\pi} \cdot \text{IrradianceCaching}(\mathbf{p}, \mathbf{n}).$$

Algorithm 5 Lazy irradiance evaluation used in irradiance caching.

    function IrradianceCaching($\mathbf{p}$, $\mathbf{n}$)
        if one or more irradiance values can be used for interpolation at $\mathbf{p}$ **then**
            return irradiance interpolated from the cached values.
        else
            Compute new irradiance value and gradients by hemisphere sampling.
            Store the value with gradients as a new record in the cache.
            return the new irradiance value.
        end if
    end function

It is advantageous to cache irradiance since it describes the amount of *incident* light at a point, before it is reflected, i.e. multiplied by local diffuse reflectivity. This allows us to reuse the cached values even on surfaces where the diffuse reflectivity varies, for example when the surface is texture mapped.

To maximize the profit of irradiance caching, we want to perform the costly hemisphere sampling only at a few locations in the scene. On the other hand, interpolation error due to irradiance reuse should be minimized. Irradiance caching addresses these two contradictory goals by selecting the locations for hemisphere sampling adaptively, such that the expected error due to interpolation is approximately constant in the whole scene.

Figure 2.8 shows the locations of records (i.e. places where hemisphere sampling was invoked) in an example scene. Notice that the records are widely spaced on flat surfaces in open areas, whereas they become more densely packed on curved surfaces and in corners. Irradiance caching always distributes the records in this way. The reason is that the expected interpolation error in the corner areas and on curved surfaces is relatively large, so irradiance caching prefers denser record spacing there.

The determination of cached irradiance values that can be used for interpolation at a given point and the interpolation procedure itself are quite intimately coupled: The interpolation is based on a simple *weighted average* of the stored values. A cached value can be used for interpolation at a point only if its interpolation weight with respect to that point exceeds some threshold value. The weight for a cached value is inversely proportional to the expected error incurred by reusing that value at the location of interpolation.

Figure 2.8: Cache record locations (black dots in the image on the right) in an example scene. The density of records in the vicinity of geometry features and on curved surfaces is automatically increased by irradiance caching in order to keep the overall interpolation error approximately constant in the whole scene. (Sponza Atrium model courtesy Marko Dabrovic.)

This scheme is designed for the very purpose of distributing the interpolation error uniformly over the whole scene and its consequence is the adaptive record spacing illustrated in Figure 2.8.

To put the above ideas into practice, we need to predict the interpolation error caused by reusing a cached value at a different location. The idea proposed by Ward et al. [WRC88] is to derive an upper bound on that error by analyzing the worst-case situation, i.e. an illumination environment that implies the largest possible error. The derivation detailed in Apendix C on page 117 uses the so called *split sphere* model as the worst case.

2.2.1   INTERPOLATION

The interpolated irradiance at $\mathbf{p}$ is computed as a weighted average of a subset of the cached irradiance values E_i:

$$E(\mathbf{p}) = \frac{\sum\limits_{i \in S(\mathbf{p})} E_i(\mathbf{p}) w_i(\mathbf{p})}{\sum\limits_{i \in S(\mathbf{p})} w_i(\mathbf{p})}, \tag{2.6}$$

where $E_i(\mathbf{p})$ is a cached irradiance value extrapolated to $\mathbf{p}$. This can simply be the cached irradiance value itself, $E_i(\mathbf{p}) = E_i$. However, translation and rotation irradiance gradients, $\nabla_r E_i$ and $\nabla_t E_i$, substantially improve extrapolation accuracy:

$$E_i(\mathbf{p}) = E_i + (\mathbf{n}_i \times \mathbf{n}) \cdot \nabla_r E_i + (\mathbf{p} - \mathbf{p}_i) \cdot \nabla_t E_i. \tag{2.7}$$

The two gradients are computed using hemisphere sampling, along with the irradiance value itself, as described in Section 2.1.

The interpolation weight $w_i(\mathbf{p})$ for the weighted average in (2.6) is defined as the inverse of the error estimate for the split sphere model (Equation C.2) derived in Appendix C:

$$w_i(\mathbf{p}) = \frac{1}{\frac{\|\mathbf{p} - \mathbf{p}_i\|}{R_i} + \sqrt{1 - \mathbf{n} \cdot \mathbf{n}_i}} - \frac{1}{a}, \tag{2.8}$$

where:

$\mathbf{p}$ is the interpolation point,
$\mathbf{n}$ is the surface normal at $\mathbf{p}$,
$\mathbf{p}_i$ is the position of the i-th cached record (stored in the cache),
$\mathbf{n}_i$ is the surface normal at $\mathbf{p}_i$ (stored in the cache),
R_i is the distance to the surfaces visible from $\mathbf{p}_i$ (computed as the harmonic mean or, alternatively, the minimum of the ray lengths in hemisphere sampling and stored in the cache),
a is a user-defined constant specifying the allowed approximation error, discussed in more detail below.

Records used in interpolation   The weight function (2.8) determines the set $S(\mathbf{p})$ of irradiance values that can contribute to interpolation at $\mathbf{p}$:

$$S(\mathbf{p}) = \{i; \ w_i(\mathbf{p}) > 0\}, \tag{2.9}$$

In other words, a cached value can be used only in the vicinity of its location, where its weight $w_i(\mathbf{p})$ is greater than zero. We will refer to this set as the *validity area* of a record. The user-defined constant a

appearing in the weight definition expresses the maximum allowed approximation error. The larger the value of a, the bigger the allowance for interpolation and the larger the interpolation error.

> *If the set $S(\mathbf{p})$ turns out to be empty at a given point, a new irradiance value is computed and stored in the cache.*

In the original formulation [WRC88], the subtraction of the $1/a$ term does not appear in the weight definition. Instead, the set of records used for interpolation is defined as $S(\mathbf{p}) = \{i; \ w_i(\mathbf{p}) > 1/a\}$. However, subtracting the $1/a$ term in the weight definition has the advantage that the weight is zero at the boundary of the validity domain, which leads to smoother interpolation results.

The irradiance records are indexed in an octree data structure that allows to quickly determine records that can be used for interpolation at a point. More details are given in Section 2.3.

In practice, an additional test is used to decide which records can be used for interpolation. Notably, we want to avoid including records from surfaces that lie "in front" of the interpolation point. Consider the situation depicted in Figure 2.9. The surface normal at $\mathbf{p}_i$ and $\mathbf{p}$ are the same and the distance to surfaces R_i for the record at $\mathbf{p}_i$ is large enough so the record i would be used for interpolation at $\mathbf{p}$. However, this would likely cause large error since the actual indirect illumination at $\mathbf{p}$ coming from the "step" next to it would be ignored. To avoid this kind of problems, Ward et al. [WRC88] suggest to compute the following quantity for each candidate record:

$$d_i(\mathbf{p}) = (\mathbf{p} - \mathbf{p}_i) \cdot (\mathbf{n} + \mathbf{n}_i)/2. \tag{2.10}$$

If $d_i(\mathbf{p})$ is less than a small negative value, then $\mathbf{p}_i$ is "in front" of $\mathbf{p}$ and the record i is excluded from interpolation.

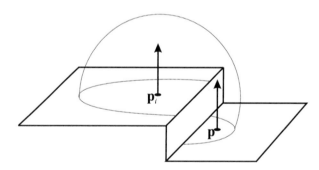

Figure 2.9: If the cached value at point $\mathbf{p}_i$ is "in front" of the interpolation point $\mathbf{p}$, it is excluded from interpolation. (Redrawn after Ward et al. [WRC88].)

Alternative Weight Formulation Tabellion and Lamorlette [TL04] use a modified version of the weight function:

$$w_i(\mathbf{p}) = 1 - \kappa \max\{\epsilon_{pi}(\mathbf{p}), \epsilon_{ni}(\mathbf{n})\}, \tag{2.11}$$

where

$$\epsilon_{pi}(\mathbf{p}) = \frac{\|\mathbf{p} - \mathbf{p}_i\|}{R_i/2} \quad \text{and} \quad \epsilon_{ni}(\mathbf{n}) = \frac{\sqrt{1 - \mathbf{n} \cdot \mathbf{n}_i}}{\sqrt{1 - \cos 10°}}. \tag{2.12}$$

The individual error terms are arbitrarily normalized to have value of 1 at the boundary of the record validity domain. (For example, normalization of the orientation error term $\sqrt{1 - \mathbf{n} \cdot \mathbf{n}_i}$ by $1/\sqrt{1 - \cos 10°}$ means that a record is never reused if the difference of $\mathbf{n}$ from $\mathbf{n}_i$ is more than 10 degrees.)

Instead of summing up the individual error terms, Tabellion and Lamorlette use their maximum in the weight definition. Using the maximum is advantageous especially for debugging since at least one of the error terms must exceed one for a record to be rejected from interpolation (as opposed to two small error terms adding up).

The user-defined constant κ determines the overall caching accuracy. It is inversely proportional to the maximum allowed interpolation error a used in the original formulation above.

Probably the most beneficial property of Tabellion and Lamorlette's weight is its overall shape, illustrated in Figure 2.10. The maximum value of 1 (right at the record location) graciously falls off to zero at the boundary of the record validity domain. Such a weight function usually produces smoother interpolation results than weight (2.8), which tends to infinity at the record location.

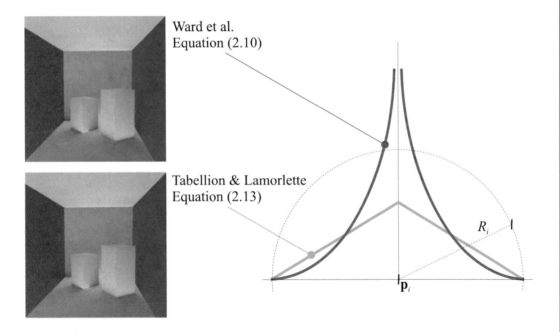

Ward et al.
Equation (2.10)

Tabellion & Lamorlette
Equation (2.13)

R_i

$\mathbf{p}_i$

Figure 2.10: The weight function (2.11) proposed by Tabellion and Lamorlette yields smoother interpolation results that the weight (2.8) of Ward et al.

2.2.2   DISTANCE TO SURFACES, R_i

The distance to the surfaces visible from the record location is calculated from ray lengths $r_{j,k}$ in hemisphere sampling when a new record is established and added to the cache. According to the derivation of the split sphere model in Appendix C, this value is computed as the *harmonic mean* of the ray lengths:

$$R_i^{\mathrm{HMD}} = \frac{MN}{\sum_{j,k} \frac{1}{r_{j,k}}}, \tag{2.13}$$

where MN is the total number of rays. As an alternative to the harmonic mean, Tabellion and Lamorlette [TL04] use the *minimum* of the ray lengths:

$$R_i^{\min} = \min_{j,k} r_{j,k}. \tag{2.14}$$

Rays near the tangent plane are not taken into account when computing the minimum, as shown in Figure 2.11. This is to avoid too small a value of $R_i^{\min}$ on slightly curved concave surfaces, where the rays near the tangent plane tend to be very short.

Figure 2.11: Rays near the tangent plane are not taken into account when computing the distance to surfaces, R_i, as the minimum of the ray lengths.

Using the minimum ray length instead of the harmonic mean has the advantage of decreasing the chance of missing some small geometric details in the interpolation.

A small sidenote: The normalizing factors in the definition of the Tabellion and Lamorlette's error terms ϵ_{pi} and ϵ_{ni} (Equation (2.12)) are optimized for the very case of defining R_i as the minimum of ray lengths.

2.2.2.1 Maximum and Minimum Record Spacing

No matter if the calculation of the distance to surfaces R_i uses the harmonic mean or the minimum of the ray lengths, the resulting value will be extremely small in corners, where many of the rays are very short. The weight computed by Equation (2.8) or (2.11) is then tiny even at a short distance between the record position $\mathbf{p}_i$ and the location of interpolation $\mathbf{p}$. As a consequence, records become overly densely packed in the corner areas, as shown in Figure 2.12 (a), and caching is not effective in there because hemisphere sampling is performed nearly everywhere. On the other hand, the value of R_i may become too big in open spaces because some of the rays in hemisphere sampling miss the surrounding geometry. The record is then reused over too large an area, which may cause some serious interpolation errors.

For the above reasons, it is a good idea to impose minimum and maximum limits on the record spacing. This can be done by clamping the R_i value computed in hemisphere sampling by some minimum and maximum thresholds, $R_{\min}$ and $R_{\max}$:

$$R_i^{\mathrm{clamp}} = \min\{\max\{R_i, R_{\min}\}, R_{\max}\}. \tag{2.15}$$

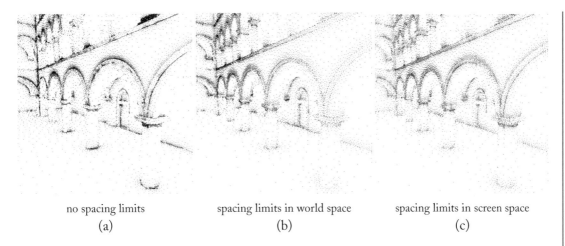

| no spacing limits | spacing limits in world space | spacing limits in screen space |
| (a) | (b) | (c) |

Figure 2.12: The effect of different strategies for imposing a minimum and maximum spacing of records by clamping the harmonic mean distance. With no spacing limits (a), too many records are generated in corner areas and caching becomes ineffective. Setting the spacing limits in world space (b) produces too many records far from the camera and too few near the camera. Setting the limits in screen space (c) yields a good distribution of records over the whole image.

One possibility is to limit the spacing in *world space*, by setting the thresholds, R_{min} and R_{max}, to a value used in the entire scene. (For example, the *ambient resolution* parameter -ar in *Radiance* [WLS98], specifies R_{min} as a fraction of the scene size; the maximum, R_{max}, is then set to 64 times the minimum.) However, limiting the record spacing in world space still produces unnecessarily densely packed records far away from the camera (see Figure 2.12 (b)), generating overkill computation. On the other hand, there will often be too few cache records near the camera, giving rise to visible interpolation problems.

A better idea, proposed in [TL04], is to clamp the record spacing relative to *screen-space* distances, by specifying R_{min} and R_{max} as a multiple of the pixel size projected to the record location. Good values for R_{min} range between 1 and 3 times the projected pixel size. The maximum of 20 times the projected pixel size is suggested in [TL04]. The projected pixel size can be computed as the square root of the pixel footprint area on the surface. In most rendering systems, the pixel footprint is readily available in the form of *ray differentials* which are used for texture filtering [PH04]. The screen-space clamping produces good distribution of records over the entire image (Figure 2.12 (c)). Moreover, it is more intuitive for users to set the spacing limits in image pixels than relative to the scene size.

2.2.2.2 Gradient-Based Limit on Record Spacing

The derivation of the interpolation criterion for irradiance caching uses the "split sphere" model as the worst case illumination scenario that implies the largest possible rate of change (or *gradient*) of irradiance with translation and thus the largest error caused by reusing an irradiance value at a different location. The gradient derived for the split sphere model is used as an upper bound on the rate of change of irradiance, which yields the weight function (2.8).

However, as discussed in Appendix C, the split sphere model represents the worst case only if the environment does not contain concentrated sources of indirect illumination. As it turns out, in some locations of many scenes, the "upper bound" derived from the split sphere model is smaller than the magnitude of the actual translation gradient $\nabla_t E_i$ estimated from hemisphere samples using Equation (2.5). The record spacing dictated by the split sphere model then becomes insufficient to properly model the illumination variation in these high-gradient areas and visible interpolation artifacts may appear.

However, we do have an estimate of the actual translation gradient at our disposal, so we can remedy the situation: The "upper bound" of the translation gradient derived from the split sphere model is given by E_i/R_i (see Appendix C). If this "upper bound" turns out to be less than the magnitude of the actual translation gradient, $\nabla_t E_i$, we simply limit the R_i value such that $\|\nabla_t E_i\| \leq \frac{E_i}{R_i}$ holds:

$$R_i \leftarrow \min\{R_i, \frac{E_i}{\|\nabla_t E_i\|}\}. \tag{2.16}$$

This heuristic is applied to a new irradiance record before inserting it into the cache. The effect of the gradient-based limit on record spacing is illustrated in Figure 2.13.

It is important, however, not to rely only on the gradient magnitude estimate $\|\nabla_t E_i\|$ to determine the record spacing since the estimate may happen to be very low. It is safer to keep the record spacing conservative by combining the predictions from the split sphere model with the actual gradient magnitude.

2.2.2.3   Limiting Gradient Magnitude

The magnitude of translation gradients calculated using Equation (2.5) may be excessively large, due to the division by the ray length $r_{j,k}$, which is often tiny along edges and in corners. Dark and bright splotches in corner areas, shown in Figure 2.14, may appear as a consequence of the exaggerated gradient magnitude.

A solution to this problem is quite straightforward. It consists in limiting the gradient magnitude as follows:

$$\nabla_t E_i \leftarrow \nabla_t E_i \cdot \min\{1, \frac{R_i}{R_{\min}}\} \tag{2.17}$$

where

R_i is the distance to surfaces calculated as mean or minimum ray length,
$R_{\min}$ is the minimum spacing threshold (see Section 2.2.2.1),
$\nabla_t E_i$ is the translational gradient (one 3D vector per color component).

The same problem is addressed slightly differently in Greg Ward's *Radiance* implementation or Eric Tabellion's PDI/Dreamworks implementation [KGW+08].

2.2.2.4   Neighbor Clamping

When a new record is created, the distance to surfaces R_i is computed from the lengths of the sample rays. However, because the sample rays do not cover all directions in the hemisphere, some geometry features in the scene may be missed. The calculated value of R_i is, then, too large and produces low record density. If the missed geometry feature is a source of strong indirect illumination, interpolation artifacts appear in the image. Because of the stochastic nature of hemisphere sampling, the features missed by one record may not be missed by another, which even amplifies the noticeability of the image artifacts.

Examples of features most commonly missed are steps of a staircase, or windowsills on a facade, which may be too small to keep the harmonic mean of ray lengths low, yet important in terms of indirect illumination. The left column of Figure 2.16 shows the artifacts due to the insufficient record density around the steps.

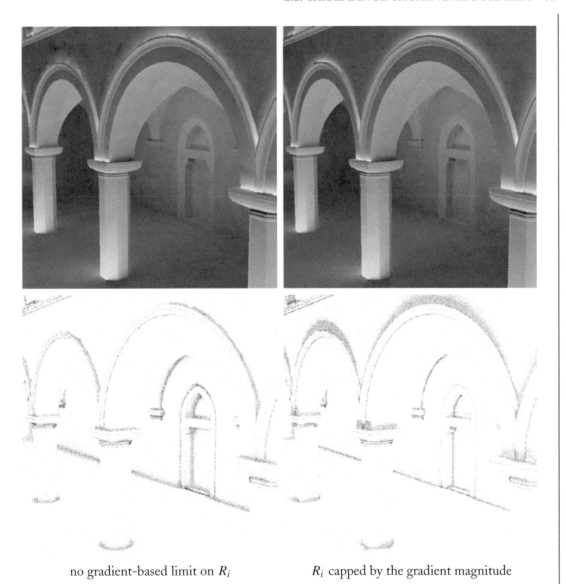

no gradient-based limit on R_i R_i capped by the gradient magnitude

Figure 2.13: The effect of limiting the distance to surfaces, R_i, by the translation gradient magnitude. The gradient of indirect illumination above the arches is very high due to the sunlight reflected from the arches onto the walls. With no gradient-based limit (left) the spacing of irradiance records predicted by the split sphere model is not sufficient to model the illumination variation. With the gradient-based limit on the distance of surfaces R_i (right), record spacing is increased around the arches and the illumination variance is properly resolved.

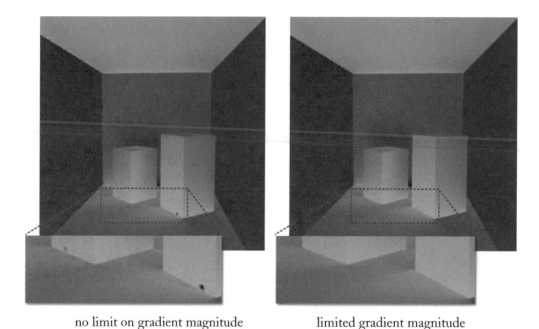

no limit on gradient magnitude   limited gradient magnitude

Figure 2.14: The effect of limiting the gradient magnitude. High gradient magnitude in corners causes image artifacts (left). The artifacts can be suppressed by limiting the gradient magnitude (right).

Tabellion and Lamorlette [TL04] address this problem by calculating the distance to surfaces R_i as a *minimum* of ray lengths, instead of taking the harmonic mean. This increases the probability of detecting small geometry features and indeed, the step-like features are less likely to be missed. However, using the minimum may be overly sensitive to even very tiny geometry features that have little significance in terms of indirect illumination.

Our solution to the problem of missing geometry features is dubbed *neighbor clamping*. We start with a simple observation: *Because of the stochastic nature of hemisphere sampling, the features missed by one record are usually not missed by all neighboring records.* Therefore, we want to propagate the information about the presence of geometry features from the records that did not miss the features onto the one that did miss them.

The second important observation is that *distance to surfaces, R_i, for neighboring records must fulfill the triangle inequality*, as illustrated in Figure 2.15. Invalidity of the triangle inequality suggests that a geometry feature was missed by some of the records.

Our solution then consists in enforcing the triangle inequality by reducing the R_i value of some records. In particular, the R_i values of two nearby records are not allowed to differ by more than the distance between the two records.

Technically, when a new record i is being created, we locate all nearby records j and clamp the new record's R_i value. After that we similarly clamp the nearby records' R_j values by the new record's R_i value. The procedure is described in Algorithm 6. The validity area of the "nearby" records overlaps with the (tentative) validity area of the new record *and* and their position passes the "in front" test (2.10).

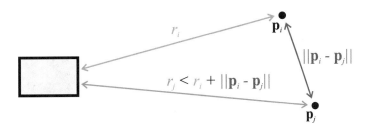

Figure 2.15: Records' distance to surfaces must fulfill the triangle inequality. Consider a record i at position $\mathbf{p}_i$ at distance r_i from a small geometry feature. Now consider another record, j, at position $\mathbf{p}_j$. By the triangle inequality, the maximum possible distance of record j from the geometry feature must be less than or equal to $r_i + \|\mathbf{p}_i - \mathbf{p}_j\|$.

Algorithm 6 Neighbor clamping heuristic.

procedure NeighborClamping($\mathbf{p}_i, \mathbf{n}_i, R_i$)
    nearbyRecords $\leftarrow$ LocateNearbyRecords($\mathbf{p}_i, \mathbf{n}_i, R_i$)

    // Clamp new record's R_i value.
    for each record j **in** nearbyRecords **do**
        $R_i \leftarrow \min\{R_i, R_j + \|\mathbf{p}_i - \mathbf{p}_j\|\}$
    end for

    // Clamp other records' R_j value.
    for each record j **in** nearbyRecords **do**
        $R_j \leftarrow \min\{R_j, R_i + \|\mathbf{p}_i - \mathbf{p}_j\|\}$.
    end for
end procedure

A consequence of this clamping is that a too large R_i value of a record, caused by missing a geometry feature, is reduced due to some of the neighboring records that did not miss that feature.

When implementing the neighbor clamping heuristic, it is important to work with the values of distance to surfaces R_i before clamping them by the minimum and maximum spacing limits $R_{\min}$ and $R_{\max}$ since only the original value contains undistorted information about the presence of geometry. However, every time a record's value of R_i is altered in neighbor clamping, we also update the value of R_i^{clamp} using Equation (2.15). For these reasons, we store with each record both the clamped and original value of distance to surfaces, R_i and R_i^{clamp}, as well as the spacing limits used for clamping, $R_{\min}$ and $R_{\max}$. (Recall that the value of $R_{\min}$ and $R_{\max}$ may be different for each record if screen-space limits are applied.)

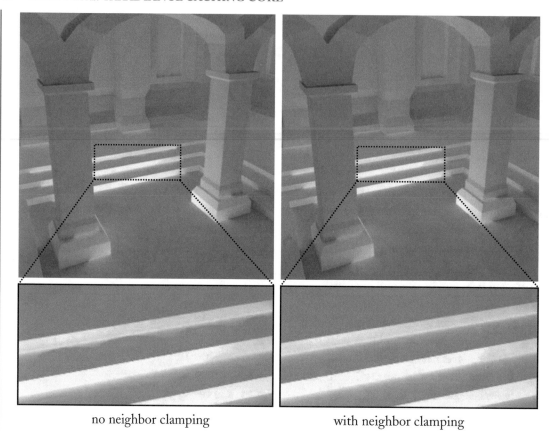

no neighbor clamping   with neighbor clamping

Figure 2.16: Without neighbor clamping, small geometry features are often missed. If the missed geometry features are sources of strong indirect illumination, image artifacts appear (see the detail of the steps in the left column). Neighbor clamping reliably detects and suppresses the artifacts caused by missing small geometry features (see the right column). The images show only indirect illumination.

The triangle inequality argument behind the neighbor clamping heuristic is fully justified only when using the minimum ray length for computing distance to surfaces R_i, but it gives very good results even for the harmonic mean. The geometry features are almost never missed, and the overall distribution of records in the scene behaves well. Figure 2.16 demonstrates how neighbor clamping (right column) helps detecting small, step-like geometry features. Without neighbor clamping (left column), those features are often missed and artifacts appear in the image (see the detail of the stairs). Both images were rendered using the same number of records (7,750). Without neighbor clamping, at least 20,000 records were required to get rid of the image artifacts on the stairs.

Ray Leaking   Irradiance caching is quite sensitive to imperfections in scene modeling; a typical example in which caching breaks down is an inaccurate connection of adjacent edges of two polygons. This may be produced *e.g.* by an insufficient number of significant digits when a scene is exported to a text file.

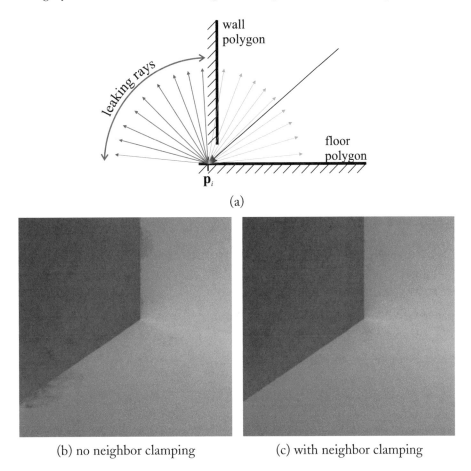

(a)

(b) no neighbor clamping   (c) with neighbor clamping

Figure 2.17: Inaccurate connection of polygons (a) may result in ray leaking, giving rise to serious image artifacts (b). The artifacts are significantly reduced by the use of the neighbor clamping heuristic (c).

Consider the situation in Figure 2.17 (a). There is a small gap between the floor polygon and the wall polygon. If a primary ray hits this gap, its intersection with the floor polygon can be found *behind* the wall polygon. As a consequence, rays that are supposed to hit the wall now *leak* either to the neighboring room or to infinity. The outcome of such an event is quite disastrous:

- The computed irradiance is completely wrong.

- The distance to surfaces R_i is much greater than it should be; therefore, the wrong irradiance value gets reused over a very large area.

An example of the resulting image artifacts is shown in Figure 2.17 (b).

Fortunately, using neighbor clamping can alleviate the situation. Records suffering from ray leaking have disproportionately greater distance to surfaces R_i than their neighbors not having this problem, thereby breaking the validity of the triangle inequality. Therefore, ray leaking is reliably detected by the use of neighbor clamping and its consequences are alleviated by the reduction of the erroneous R_i value. Figure 2.17 (c) shows that neighbor clamping detects and suppresses the effects of ray leaking.

2.2.3   CREATING AN IRRADIANCE RECORD: SUMMARY

In the previous section, we have seen that a number of heuristics may have impact on the values stored in the cache. Let us now summarize the whole process of creating a new irradiance record:

Algorithm 7 Procedure used to create a new irradiance cache record.

function CreateRecord($\mathbf{p}_i$, $\mathbf{n}_i$, i)

    $[L_{j,k}]$, $[r_{j,k}] \leftarrow$ SampleHemisphere($\mathbf{p}_i$, $\mathbf{n}_i$)

    $E_i \leftarrow$ IrradEstimate($[L_{j,k}]$)   ▷ *Equation (2.2), pg. 19*

    $\nabla_r E_i \leftarrow$ RotGrad($[L_{j,k}]$)   ▷ *Equation (2.4), pg. 21*

    $\nabla_t E_i \leftarrow$ TransGrad($[L_{j,k}]$, $[r_{j,k}]$)   ▷ *Equation (2.5), pg. 23*

    // A series of heuristics sets the value of R_i...

    $R_i \leftarrow$ DistanceToSurfaces($[r_{j,k}]$)   ▷ *Equation (2.13) or (2.14), pg. 30*

    Limit R_i by gradient: $R_i \leftarrow \min\{R_i, \frac{E_i}{\|\nabla_t E_i\|}\}$ ▷ *Equation (2.16), pg. 32*

    Clamp R_i: $R_i^{\text{clamp}} = \min\{\max\{R_i, R_{\min}\}, R_{\max}\}$ ▷ *Equation (2.15), pg. 30*

    Limit gradient by R_i: $\nabla_t E_i \leftarrow \nabla_t E_i \cdot \min\{1, \frac{R_i}{R_{\min}}\}$ ▷ *Equation (2.17), pg. 32*

    NeighborClamping($\mathbf{p}_i$, $\mathbf{n}_i$, R_i, R_i^{clamp}, $R_{\min}$, $R_{\max}$)   ▷ *Algorithm 6, pg. 35*

    record $\leftarrow$ AllocateRecord($\mathbf{p}_i$, $\mathbf{n}_i$, E_i, $\nabla_r E_i$, $\nabla_t E_i$, R_i^{clamp}, R_i, $R_{\min}$, $R_{\max}$)

    return record

end function

An irradiance record contains the following entries:

$\mathbf{p}_i$	Point3D	record location
$\mathbf{n}_i$	Vector3D	surface normal at $\mathbf{p}_i$
E_i	Color	cached irradiance
$\nabla_r E_i$	Vector3D[]	rotation gradient (one 3D vector per color component)
$\nabla_t E_i$	Vector3D[]	translation gradient (one 3D vector per color component)
R_i^{clamp}	float	distance to surfaces used for irradiance interpolation
R_i	float	original distance to surfaces (used for neighbor clamping)
$R_{\min}$	float	minimum spacing threshold (used for neighbor clamping)
$R_{\max}$	float	maximum spacing threshold (used for neighbor clamping)

2.3   DATA STRUCTURE

In order to make irradiance caching effective, there should not be much overhead related to the cache queries. For that we need a storage method for the cached values that ensures fast lookups. We definitely cannot afford to evaluate the weight function for all the cached records every time we determine the set $S(\mathbf{p})$ of records usable for interpolation at a point.

One option would be to store the irradiance values directly on the surface elements of the geometry, such as triangle faces, in a way similar to radiosity algorithms [CW93]. This approach would afford for very fast lookups but it has several important disadvantages: the geometry would be limited to parametric patches, stored irradiance values could not be reused over several objects, there would be problems with instancing of geometry, etc. A better option is to keep the record storage completely independent of the scene representation in a spatial index structure.

From the definition of the set $S(\mathbf{p})$ (Equation 2.9) and the weight $w_i(\mathbf{p})$ (Equation 2.8), we see that a record can only be reused inside the *validity sphere* centered at the record location with the *validity radius* of aR_i. Therefore, we are looking for a spatial data structure for indexing spheres of various radii that allows efficient insertion (without a complete rebuild of the structure) and, more importantly, fast lookup of all the spheres that overlap a given point in space.

2.3.1   SINGLE-REFERENCE OCTREE

Ward at al. [WRC88] recommend to organize the records in an octree. The octree used for the irradiance cache is completely independent of the data structure used to accelerate ray-object intersections. Each record is referenced by a single octree node—the node that contains the record location and the box size of which is proportional to the record's validity radius. The octree is built incrementally as new records are created and inserted. The insertion proceeds from the root to the node that will reference the new record, possibly creating new nodes on the way. Figure 2.18 (a) shows an example of the assignment of records to the octree nodes. The procedure to look up records that can contribute to the interpolated value at point $\mathbf{p}$ is summarized in Algorithm 8.

Algorithm 8 Record lookup in the single-reference octree.

---

    procedure LookUpRecordsSR(node, $\mathbf{p}$, $\mathbf{n}$)
        for all records i stored in node **do**   ▷ *Examine records in this node*
            if $(w_i(\mathbf{p}) > 0)$ **and** ($\mathbf{p}_i$ not in front of $\mathbf{p}$) **then**
                Include record in $S(\mathbf{p})$.
            end if
        end for
        for all children of node **do**   ▷ *Recurse*
            if $\mathbf{p}$ is within half the child's size of its cube boundary **then**
                LookUpRecordsSR(child, $\mathbf{p}$, $\mathbf{n}$)
            end if
        end for
    end procedure

This algorithm searches not only the octree nodes containing the point $\mathbf{p}$, but any octree node with a boundary within half the node's side length of $\mathbf{p}$. This guarantees that all relevant cached records

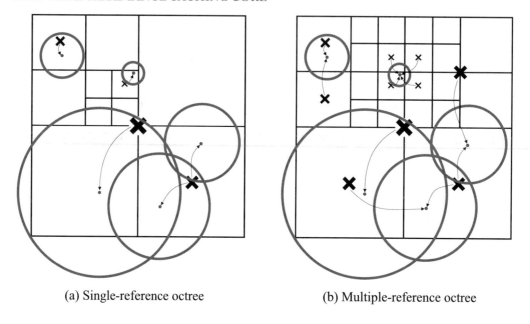

(a) Single-reference octree

(b) Multiple-reference octree

Figure 2.18: Example of the assignment of irradiance cache records to the octree nodes for the single-reference octree (a) and the multiple-reference octree (b). In the single-reference octree, a record is referenced by a single octree node containing the record's location. In the multiple-reference octree, on the other hand, a record is referenced by all the tree nodes overlapped by the record's validity sphere (shown here as the thick circles). (Figure adapted from [WLS98].)

will be examined. The single-reference octree is used in the *Radiance* implementation of irradiance caching [WLS98].

2.3.2   MULTIPLE-REFERENCE OCTREE

The disadvantage of the single-reference octree described above is the relatively slow search procedure that involves possible recursive examination of multiple children at each node. Since our aim is to cut the cache query time to the minimum, it is a good idea to design the traversal procedure such that the recursion can be eliminated. This can be easily achieved by referencing each record from all octree nodes that overlap the record's validity sphere, as illustrated in Figure 2.18 (b). The lookup procedure now merely examines the records stored in the nodes on the way from the root to the leaf containing the query point (Algorithm 9).

To *insert* a new record in the octree, we recursively traverse the octree nodes that overlap the record's validity sphere (creating new nodes on the fly as needed). The traversal is stopped and a reference to the new record is created when the node's box is approximately the same size as the validity radius of the new record.

The disadvantage of the multiple-reference octree are an increased memory consumption and slower record insertion. However, they are more than compensated by the much faster lookup.

Algorithm 9 Record lookup in the multiple-reference octree.

---

    procedure LookUpRecordsMR($\mathbf{p}$, $\mathbf{n}$)
        node $\leftarrow$ root
        while node $\neq$ NULL **do**
            for all records i stored in node **do**   ▷ *Examine records in this node*
                if ($w_i(\mathbf{p}) > 0$) **and** ($\mathbf{p}_i$ not in front of $\mathbf{p}$) **then**
                    Include record in $S(\mathbf{p})$.
                end if
            end for
            node $\leftarrow$ child containing $\mathbf{p}$ ▷ *Descend in the octree*
        end while
    end procedure

A basis for the implementation of the multiple-reference octree can be found in PBRT [PH04] (**octree.h**). Algorithm 10 gives a C++ code for a sphere-box overlap test [LAML07].

Algorithm 10 C++ code for a sphere-box overlap test.

```cpp
// Return true if a sphere (c,r) overlaps a box given by corners pMin and pMax.
bool SphereBoxOverlap(const Vector3D& pMin, const Vector3D& pMax,
                      const Vector3D& c, const float r) const
{
  float dmin = 0;
  for( int i = 0; i < 3; i++ )
  {
    if      ( c[i] < pMin[i] ) dmin += sqr(c[i] - pMin[i]);
    else if ( c[i] > pMax[i] ) dmin += sqr(c[i] - pMax[i]);
  }
  return dmin <= r*r;
}
```

2.3.3   LAST QUERY REUSE

When super-sampling is used for image anti-aliasing, there may be many irradiance cache queries issued for shading the same pixel. In most cases, these queries will return very similar irradiance values, which are in turn averaged in the calculation of the final pixel color. It is a waste of time to run all these cache queries since reusing the results of a query for all subsequent queries in the same pixel produces nearly identical results. However, we must ensure that a query result is not reused over discontinuities in depth and surface orientation. This can be accomplished by checking the following conditions before the result of the last query is reused:

- The last query was issued for computing the value of the same pixel as the new query.

- The distance between the locations of the two queries is less than the projected pixel size.

- The difference in normals is less than a threshold (we use $\mathbf{n}_{last} \cdot \mathbf{n}_{new} > 0.95$).

If any of the above conditions is not met, we issue a new cache query, instead of reusing the last value.

Even for moderate super-sampling rates, the savings due to the last query reuse may be substantial. In addition, this simple technique also amortizes the penalty for a not very well optimized implementation of the spatial data structure used for record look-ups.

2.4   IRRADIANCE CACHING SUMMARY

What we have described by now should be sufficient to implement the "guts" of irradiance caching. Before going to the next section, which focuses on the integration of irradiance caching in a renderer, let us summarize the algorithm. We take the brief description of irradiance caching from Algorithm 5 on page 25 and fill in the references to various sub-procedures described earlier. The result is shown in Algorithm 11.

Algorithm 11 A detailed overview of the irradiance caching algorithm.

function IrradianceCaching($\mathbf{p}$, $\mathbf{n}$)
    $S(\mathbf{p}) \leftarrow$ LookUpRecordsMR($\mathbf{p}$, $\mathbf{n}$) $\triangleright$ *Algorithm 9, pg. 41*
    if $S(\mathbf{p}) \neq \emptyset$ **then**
        return InterpolateIrradiance($S(\mathbf{p})$, $\mathbf{p}$, $\mathbf{n}$) $\triangleright$ *Equation 2.6, pg. 27*
    else
        record $\leftarrow$ CreateRecord($\mathbf{p}$, $\mathbf{n}$, recordCount) $\triangleright$ *Algorithm 7, pg. 38*
        StoreRecord(record) $\triangleright$ *Section 2.3, pg. 39*
        return record.E
    end if
end function

CHAPTER 3

Practical Rendering with Irradiance Caching

This chapter discusses some details that make rendering with irradiance caching more practical. Section 3.1 shows that the order in which pixels are visited during image rendering may have a significant impact on the caching performance and image quality. In Section 3.2 we focus on maintaining a good performance of irradiance caching in complex scenes. After that, in Section 3.3, we turn our attention to motion blur rendering. Finally, Section 3.4 shows how irradiance caching can be used for ambient occlusion computation.

3.1 SINGLE-PASS VS. TWO-PASS IMAGE RENDERING

One of the strong features of irradiance caching is the "lazy evaluation" scheme: new irradiance values are computed on the fly, only if the existing cached values cannot be used for interpolation at a given point. However, the lazy evaluation may compromise the image quality depending on the order in which pixels are visited during image rendering.

3.1.1 SCANLINE ORDER

Consider image rendering in scanline order, one pixel after another, from the top left to the bottom right. In such a scenario, a record created on a scanline will not affect the color of the preceding scanlines at all, even though some pixels on the preceding scanlines may fall within the validity area of the new record. This may, and usually does, create disturbing interpolation artifacts in images, shown in Figure 3.1 (a). For brevity, we will refer to this phenomenon as the "*missed contribution*" problem.

3.1.2 HIERARCHICAL REFINEMENT AND BEST CANDIDATE PATTERN

The situation is greatly improved by using a hierarchical refinement of the image. First, only several pixel on several scanlines are shaded, effectively producing a very low-resolution version of the final image. Each subsequent pass quadruples the number of shaded pixels by doubling the effective image resolution along both axes.

Similar to the scanline order, hierarchical refinement does suffer from the "missed contribution" problem: once shaded, some pixels are not affected by irradiance records created at a later time, even if they fall within the new records' validity areas (the strange patterns in Figure 3.1 (b) are caused by this problem). Still, hierarchical refinement greatly ameliorates the situation for the following reasons:

(1) fewer pixel in the image suffer from the "missed contribution" problem,

(2) erroneous pixels do not form contiguous blocks, which makes the errors less noticeable.

In addition, the number of records required to cover the whole image is lower than with the scanline order (a consequence of point (1) above), yielding better overall rendering performance.

Using the Best Candidate pattern [PH04, Section 7.5] for image sampling takes the aforementioned advantages of hierarchical refinement even further: even fewer pixels suffer from "missed contribution"; even fewer records are required to cover the image plane; the erroneous pixels form even less regular, and less conspicuous, patterns.

The Best Candidate pattern is created from a single randomly placed point. In the subsequent steps, several randomly placed candidate points are proposed, and the one farthest from all existing points is kept. The procedure continues until a desired number of points is created.

3.1.3   TWO-PASS RENDERING

No matter how good the image sampling pattern, it is impossible to completely suppress the "missed contribution" problem, described above, in a single-pass rendering. It is often unavoidable to resort to a two-pass rendering approach. In the first pass, we use hierarchical refinement or Best Candidate pattern to sample the image, with the only goal to populate the irradiance cache. In the second pass, we use the cached irradiance values in shading image pixels. The pixel traversal order in the second pass can be arbitrary since the cache has been filled in the first pass. To reduce the overhead due to the two passes, it is not necessary to generate a fully antialiased image in the first-pass. However, it is a good idea to generate one sample per pixel if fine illumination details are to be resolved properly. Figure 3.1 (c) shows the result of two-pass rendering. Notice that the pixel patterns appearing in (b) due to the missed contributions are now gone.

Smoothing out indirect illumination   The two-pass rendering approach allows us to use another trick to improve the smoothness of the interpolated illumination. All it takes is to increase the value of the allowed interpolation error a in the second pass (see Section 2.2.1, pg. 27). Increasing a effectively enlarges the validity domain of each record and produces smoothing of indirect illumination, as shown in Figure 3.1 (d). Smoothing introduces more bias in the result but on the other hand, makes the error due to interpolation much less conspicuous. In addition, smoothing also reduces flickering in animation rendering. Typically, we increase the a value in the second pass from $1.4\times$ to $1.8\times$.

3.2   HANDLING COMPLEXITY

The efficiency of irradiance caching relies on the validity of the technique's fundamental assumption that indirect illumination varies slowly in space. This is why using irradiance caching on complex geometry, such as trees, will rarely pay off—simple path tracing usually does a better job in complex scenes.

Some of the problems caused by complex geometry can be alleviated by using a simplified scene geometry for irradiance caching, as described in Section 3.2.1. In addition, in some cases we know à priori the character or the source of the scene complexity. In such cases, we can exploit our knowledge to get the best of irradiance caching even in complex scenes. In the following sections, we will discuss some of these cases, more specifically bump mapping, displacement mapping, and rendering of fur, hair and grass.

3.2.1   RAY TRACING SIMPLIFIED GEOMETRY

When performing hemisphere sampling for the purpose of creating a new irradiance record, the accuracy of individual radiance samples computed using ray tracing is not critical since they are anyhow averaged into just a single value, the irradiance E. We can take advantage of this by ray tracing only a coarse version of the scene geometry in irradiance caching (we select only a coarse subdivision or tesselation level for

(a) scanline order   (b) hierarchical refinement

(c) two-pass rendering   (d) two-pass rendering with 1.6× smoothing

Figure 3.1: The impact of pixel rendering order on the image quality. Scanline order (a) produces discontinuities. Hierarchical image refinement (b) improves the image quality but still leaves artifacts in the form of pixel patterns. Two-pass rendering (c) suppresses the patterns. Additional smoothing (d) removes the visible discontinuities due to interpolation. (No irradiance gradients were used to make image artifacts more apparent. With gradients the differences are more subtle.)

curved surfaces). Using simplified geometry makes ray casting[1] cheaper and avoids problems with scenes that would not fit into memory in full detail.

To make ray tracing of simplified geometry work, one must resolve the inconsistency between the full-detail geometry and the simplified one: Rays are initiated from a point on the detailed geometry but

---

[1] By *ray casting* we understand finding the nearest intersection for a given ray.

only the simplified one is used to find intersections. This discrepancy may lead to self-intersections, when an unwanted intersection is reported with the surface from which the ray was initiated. The solution suggested by Tabellion and Lamorlette [TL04] is illustrated in Figure 3.2. They use the following ray offsetting algorithm:

1. Find all ray-geometry intersections within a user-defined offset distance from the ray origin (on both sides of the ray origin).

2. Stop ray traversal once an intersection is reported beyond the offset distance.

3. Let the nearest of the intersections found within the offset distance become the new effective ray origin. If no intersection is found within the offset distance, leave the ray origin unchanged.

4. Return the next hit along the ray as the resulting ray-geometry intersection.

   For this algorithm to work, the offset distance should be sightly greater than the maximum distance between the original and coarse geometry.

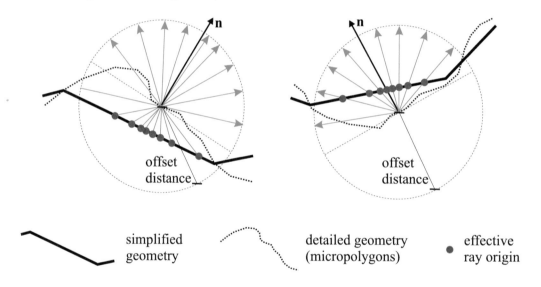

Figure 3.2: Ray offsetting algorithm used for ray tracing of simplified geometry. (Redrawn after [TL04].)

   The global illumination system described by Tabellion and Lamorlette [TL04] uses manual setting for the geometry tesselation rate. Christensen el al. [CLF+03] describes an automatic approach to set the tesselation rate for ray tracing depending on ray differentials.

3.2.2   BUMP MAPPING

In bump mapping, a texture modulates the surface normal of a base surface, creating the appearance of a bumpy or wrinkled surface. The normal variation due to the bump map can be substantial. Since irradiance caching reuses values only over areas with similar normals, its performance significantly degrades on a bumped surface—many records are created due to the high normal variation, as shown in Figure 3.4 (a).

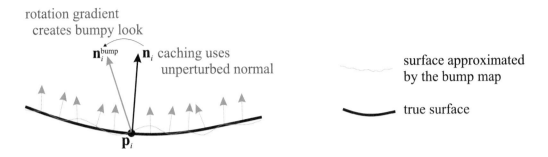

Figure 3.3: Irradiance caching on a bump mapped surface uses the unperturbed normal in order to maintain good cache coherency. The bumped look of indirect illumination is approximated by applying the rotation gradient to extrapolate irradiance to the surface orientation given by the bump map.

A simple solution to this problem, illustrated in Figure 3.3, consists in using the surface normal unperturbed by the bump map for hemisphere sampling and the cache queries. To reintroduce the detail due to the bump map, we use the rotation gradient to extrapolate irradiance to the surface orientation given by the bumped normal. The result is shown in Figure 3.4 (c). Although some detail of the bump map is lost, the overall appearance of the bumpy surface is well preserved.

3.2.3   DISPLACEMENT MAPPING

Displacement mapping is similar to bump mapping with the difference that not only the surface normal, but also the surface position itself is modulated by the texture. If irradiance caching was simply applied to the displaced surface's position and orientation, the limited degree of coherence would cause poor caching performance.

To work out this problem, Tabellion [KGW+08] suggests an approach similar to what we used for bump mapping (see Figure 3.5):

- Hemisphere sampling uses displaced surface but unperturbed normal. (The displaced position is used to avoid self-intersection problems.)

- Cache records are stored on the undisplaced surface.

- Cache lookups are performed on the undisplaced surface.

Translation and rotation gradients are used to adjust irradiance when moving from one position and orientation to another.

As a consequence of the above technique, the illumination details due to indirect light reflected from one displacement onto another are lost. To compensate for this loss, it is usually sufficient to modulate the indirect illumination by ambient occlusion (see Section 3.4 below), statically pre-computed for the displacement map and mapped onto the displaced surface.

3.2.4   FUR, HAIR & GRASS

Fur, hair and grass present a challenge for irradiance caching due to their sheer geometric complexity. Fortunately, if visual plausibility matters more than accuracy, a simple solution due to Tabellion and

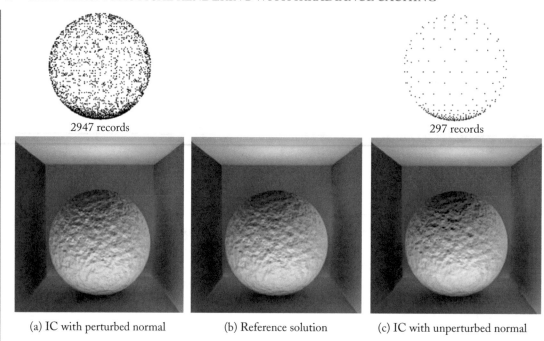

2947 records   297 records

(a) IC with perturbed normal   (b) Reference solution   (c) IC with unperturbed normal

Figure 3.4: Example of irradiance caching on a bump mapped surface. a) Using the perturbed normal vector directly in irradiance caching yields poor cache coherency, and as a consequence, many records are generated. c) With the unperturbed normal, the number of irradiance cache records drops tenfold in our example, although some bump detail is lost. The top row shows the record positions. The renderings in the bottom row show indirect illumination only.

Lamorlette [TL04] can be used to apply irradiance caching to furry characters or grassy grounds. The irradiance queries are performed on the base surface, at a position corresponding to the root of the hair or grass blade. Hair is completely ignored in hemisphere sampling when a new record is created. That is to say, there is no hair or grass present in the scene from the point of view of irradiance caching.

Irradiance interpolation is a two-step process. First, the irradiance is interpolated at the root of the shaded hair using translation and rotation gradient as in normal irradiance caching. In the second step, the interpolated irradiance at the hair root is extrapolated along the hair using a fake self-shadowing gradient based on the distance along the hair [NK04].

As a consequence of this approach, hair receives indirect illumination from distant surfaces but not from the underlying character skin or from other hair. This may be a problem if accuracy was important but for stylized depiction in computer animation this is usually not an issue. It should be pointed out, however, that the above approach is not suitable for computing indirect illumination on long hair.

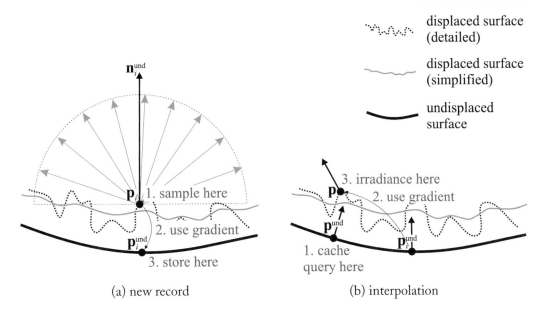

displaced surface
(detailed)

displaced surface
(simplified)

undisplaced
surface

$\mathbf{n}_i^{\mathrm{und}}$

$\mathbf{p}_i$ 1. sample here

2. use gradient

$\mathbf{p}_i^{\mathrm{und}}$

3. store here

(a) new record

3. irradiance here

$\mathbf{p}$ 2. use gradient

$\mathbf{p}^{\mathrm{und}}$

$\mathbf{p}_i^{\mathrm{und}}$

1. cache
query here

(b) interpolation

Figure 3.5: Irradiance caching and displacement mapping. (a) To create a new radiance record, hemisphere sampling uses the displaced surface and unperturbed normal. However, when stored in the cache, record position uses the corresponding point on the undisplaced surface. Translation gradient is applied to adjust irradiance. (b) To interpolate irradiance at point $\mathbf{p}$ on the displaced surface, the cache query is performed at the corresponding undisplaced point $\mathbf{p}^{\mathrm{und}}$. However, the irradiance is extrapolated to the original point, $\mathbf{p}$.

3.3   MOTION BLUR

Motion blur is widely used in photography or cinematography to convey the speed of a moving object. It is thus utterly important to make irradiance caching compatible with motion blur rendering.

Motion blur is essentially a low-pass filter that blurs parts of the image. This is why it is often applied in image post process, after the image has been rendered. However, for accurate results, motion blur should be generated as an integral part of image rendering.

In micropolygon-based architectures, such as Reyes [CCC87], used in most implementations of the RenderMan standard [Ups90], the micro-polygon vertices are shaded before the visibility and the motion blur are resolved. In other words, the shading calculation is not aware of motion blur, and therefore, it proceeds the same, no matter if motion blur is used or not. This makes the application of irradiance caching straightforward.

However, in rendering systems based purely on ray tracing (or any other rendering technique where visibility determination is not decoupled from shading), the situation is more complicated. Let us first recall the distribution ray tracing approach to motion blur rendering [CPC84, Coo89]. To compute the color of a pixel, a distribution ray tracer generates a number of rays passing through random positions inside the pixel. For each of these rays, a random time in the "shutter open" time span is selected and the

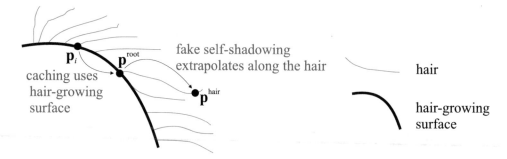

Figure 3.6: Irradiance caching on hair and fur essentially ignores the presence of hair. Cache queries are performed on the base surface at the hair root. The irradiance at the root is then extrapolated to the position on the hair using a fake self-shadowing term. This approach works well for short and medium hair, fur or grass.

ray is traced in the scene, where geometry position is adjusted to that time. Algorithm 12 summarizes this procedure.

Algorithm 12 Motion blur computation with distribution ray tracing.

    function ComputePixelRadiance
        $L_{\text{out}} \leftarrow 0$
        for $i \leftarrow 1$ **to** samplesPerPixel **do**
            $t_i \leftarrow$ random time in the "shutter open" time span
            $\text{dir}_i \leftarrow$ ray direction through a random location in the pixel
            $\text{ray}_i \leftarrow [\text{cameraPos}, \text{dir}_i]$
            L_{out} += $\text{Trace}(\text{ray}_i, t_i)$ ▷ *Adjust geometry to time t_i*
        end for
        return $L_{\text{out}}/$ samplesPerPixel
    end function

In short, if there are moving objects in the scene, each single primary ray in distribution ray tracing essentially traces a (slightly) different scene. If irradiance caching was simply applied at each intersection without additional care, we would obtain quite disastrous results, shown in Figure 3.7(a). The reason is twofold:

- The cache query points are distributed all along the trajectory of the moving objects. However, irradiance caching expects the query points on a smooth surface. Therefore, there will be an excessive number of cache records.

- Each irradiance value in the cache corresponds to a different time. Interpolation between these values does not make any sense.

One possible solution is to use "temporal re-projection." Whenever we are about to perform an irradiance cache query, we first move the point and its normal to a location corresponding to one fixed time,

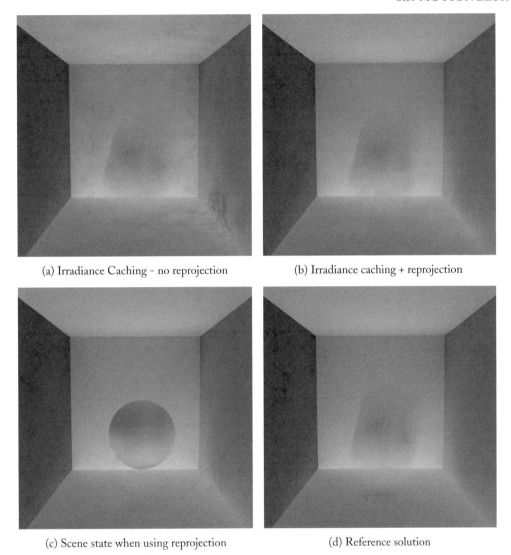

(a) Irradiance Caching - no reprojection

(b) Irradiance caching + reprojection

(c) Scene state when using reprojection

(d) Reference solution

Figure 3.7: (a) Using irradiance caching to render a scene with motion blur produces image artifacts since each cache record captures illumination at a different instant in time. (b) A simple solution to solve this problem is to "reproject" the query points to the corresponding position at a fixed instant in the shutter open time span. (c) When using the temporal reprojection, irradiance caching sees the scene in the state corresponding to that fixed time. (d) The reference solution may differ from the irradiance caching solution (notice the darker spot on the floor under the sphere) since indirect illumination is not motion blurred when using the temporal reprojection. (The images show indirect illumination only.)

typically in the middle of the shutter open time span. Therefore, all irradiance calculations correspond to the same scene configuration and irradiance can be interpolated. The disadvantage of this approach is that the indirect illumination is not motion blurred. However, this is rarely a problem, since global illumination usually changes very slowly in time.

3.4   AMBIENT OCCLUSION CACHING

Ambient occlusion is a scalar value between zero and one associated with positions on a surface. It expresses what portion of the hemisphere above each surface point is shadowed by the scene geometry.

To apply ambient occlusion in rendering, we simply multiply the results of a local shading, such as diffuse or even constant, by the ambient occlusion value. As a result, surfaces surrounded by a lot of geometry appear dark whereas surfaces that are not occluded appear light, which is similar to how an object would appear on a cloudy day (see Figure 3.9). Ambient occlusion has recently become very popular both for real-time and in high-quality rendering [Lan02, Chr03], since it can render some aspect of global illumination—the soft illumination gradations—at a much lower cost.

Ambient occlusion $A(\mathbf{p})$ can be formally defined as:

$$A(\mathbf{p}) = \frac{1}{\pi} \int_{H^+} V(\mathbf{p}, \omega) \cos \theta \, d\omega \tag{3.1}$$

where $\mathbf{p}$ is the considered point on a surface, H^+ is the upper hemisphere centered around the surface normal at $\mathbf{p}$, ω is a direction, θ is the angle between of ω and $\mathbf{n}$, the surface normal at $\mathbf{p}$, i.e. $\cos \theta = \mathbf{n} \cdot \omega$. The visibility function $V(\mathbf{p}, \omega)$ returns one if $\mathbf{p}$ is visible in the direction ω and zero otherwise (see Figure 3.8). The normalization by $1/\pi$ ensures that $A(\mathbf{p}) = 1$ in the case of no occlusion. The above

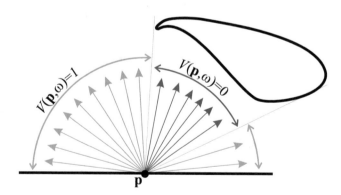

Figure 3.8: Ambient occlusion expresses what portion of the hemisphere above a point is occluded by the scene geometry. It is computed as a hemispherical integral of the visibility function V.

definition shows the relation of ambient occlusion to irradiance $E(\mathbf{p})$, defined by Equation (2.1), pg. 18 as:

$$E(\mathbf{p}) = \int_{H^+} L_i(\mathbf{p}, \omega) \cos \theta \, d\omega.$$

Ambient occlusion is nothing more than irradiance on a surface illuminated by infinitely distant light source with unit radiance L_i in all directions:

$$A(\mathbf{p}) = \frac{1}{\pi} E(\mathbf{p}) \Big|_{\forall \omega:\ L_i(\mathbf{p}, \omega) = 1}.$$

To sum up, ambient occlusion is a special case of irradiance. Therefore, all the observations and assumptions we made in deriving irradiance caching also hold for ambient occlusion, and consequently we can use irradiance caching to speed up the computation of occlusion values. The only difference is in hemisphere sampling. While in irradiance estimation, each sample ray brought a radiance sample $L_{j,k}$, in ambient occlusion estimation, each sample ray brings a visibility sample—either zero or one, depending on whether the ray hit an object or not. The rest of the algorithm, including the gradient calculation remains the same. All we do is replace all radiance samples with the samples of visibility.

(a) Ambient occlusion

(b) Locations of ambient occlusion calculation

(c) Constant shading

(d) Ambient occlusion · constant shading

Figure 3.9: Ambient occlusion (a) expresses the portion of the hemisphere above each surface point that is occluded by other surface. The calculation of ambient occlusion values can be accelerated by irradiance caching (b). Ambient occlusion is used to modulate the results of the local shading model, for example the trivial constant shading shown in (c), to produce a more realistic image (d). (Model courtesy of Universal Production Partners.)

CHAPTER 4

Irradiance Caching in a Complete Global Illumination Solution

In this chapter, we treat the irradiance caching algorithm as a black box and we focus on its efficient integration in the process of computing full global illumination solution.

Computing full global illumination in a scene involves solving the Rendering Equation 1.5 (pg. 16). Using recursive ray tracing, the Rendering Equation can be solved through recursive evaluation of the *illumination integral*:

$$L_o(\mathbf{p}, \omega_o) = L_e(\mathbf{p}, \omega_o) + \int_{H^+} L_i(\mathbf{p}, \omega_i) f_r(\mathbf{p}, \omega_i, \omega_o) \cos \theta_i \, d\omega_i,$$

which expresses total outgoing radiance $L_o(\mathbf{p}, \omega_o)$ at point $\mathbf{p}$ in direction ω_o as a sum of self-emission $L_e(\mathbf{p}, \omega_o)$ and reflected radiance $L_r(\mathbf{p}, \omega_o)$, given by a hemispherical integral of the product of incoming radiance L_i, the BRDF f_r, and the cosine term $\cos \theta_i$.

Evaluating the illumination integral at a point involves obtaining the samples of incoming radiance through ray tracing. These samples are, in turn, computed by evaluating the exact same integral, this time at a different location in the scene (see Figure 4.1). Thus the ray tracing's recursive nature.

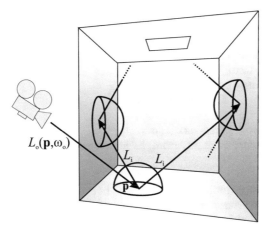

Figure 4.1: Solution of the rendering equation through recursive evaluation of the illumination integral.

In image rendering, our end goal is to evaluate the illumination integral for all surface points $\mathbf{p}$ visible through image pixels, with the outgoing direction ω_o aiming towards the camera location.

4.1   PATH GRAMMAR

Global illumination considers any possible light path from the light sources to a given point in the scene, and eventually the eye. We use the path grammar proposed by Heckbert [Hec90] to identify the light paths. Path grammar is a regular grammar consisting of symbols representing the following events:

L - light emission from the light source
D - reflection on a diffuse surface
S - reflection on a glossy/specular surface
E - light reaches the eye (i.e. the virtual camera)

A regular expression identifies a class of paths. The expressions are formed using the common notation:

ab - concatenation of a and b
$a|b$ - union (either a or b)
a^* - zero or more repetition of a
a^+ - one or more repetition of a

For example, any light path from the light to the eye is given by the regular expression $L(D|S)^*E$. Direct illumination is given by $L(D|S)E$, since it is due to light that reflects exactly once on its way from the light source to the eye.

4.2   ILLUMINATION COMPONENTS

Similar to Jensen [Jen01], we split the BRDF into three major components, each of which tells us how light is reflected at the considered point:

$f_{r,d}$ ideal diffuse (or Lambertian) term,
$f_{r,s}$ ideal specular (or mirror) term,
$f_{r,g}$ glossy (or directional diffuse) term.

In addition, we split the incoming radiance L_i into the following components, each describing what interactions light has undergone on the way from the light source to the considered point:

$L_{i,l}$ direct illumination (radiance coming directly from light sources),
$L_{i,c}$ caustics (radiance after one or more specular/glossy interactions with no diffuse reflection),
$L_{i,d}$ smooth indirect diffuse illumination (radiance after one or more diffuse reflections).

For the purpose of rendering, we can now write the illumination integral as a sum of several components, each of which is solved by a specialized technique. We omit the arguments $\mathbf{p}$, ω_i, and ω_o for brevity.

$$L_o = \quad L_e+ \qquad\qquad\qquad\qquad \bullet \text{ self-emission, } LE \text{ paths}$$

$$\int L_{i,l}(f_{r,d} + f_{r,g})\cos\theta_i \, d\omega_i+ \qquad \bullet \text{ direct illumination, } L(D|S)E \text{ paths}$$

$$\int L_i f_{r,s} \cos\theta_i \, d\omega_i+ \qquad\qquad \bullet \text{ ideal specular reflection, } L(D|S)SE \text{ paths}$$

$$\int (L_{i,c} + L_{i,d}) f_{r,g} \cos\theta_i \, d\omega_i+ \qquad \bullet \text{ glossy indirect, } L(D|S)SE \text{ paths}$$

$$\int L_{i,c} f_{r,d} \cos\theta_i \, d\omega_i+ \qquad\qquad \bullet \text{ caustics, } LS^+DE \text{ paths}$$

$$\int L_{i,d} f_{r,d} \cos\theta_i \, d\omega_i \qquad\qquad \bullet \begin{array}{l} \text{smooth diffuse indirect, } L(D|S)DDE \text{ paths} \\ \text{(computed with } \textit{irradiance caching}) \end{array}$$

Figure 4.2 shows a tabular organization of these illumination components.

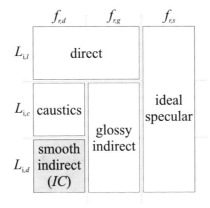

Figure 4.2: Tabular organization of illumination components. Irradiance caching (IC) is used only for the calculation of the smooth diffuse indirect illumination.

4.2.1 SMOOTH DIFFUSE INDIRECT ILLUMINATION

The only illumination component handled by irradiance caching is the smooth indirect diffuse illumination (or *diffuse interreflections*) (Figure 4.3 a). In practice, it means that:

(a) Irradiance caching only handles the ideal diffuse (Lambertian) component of the BRDF at the point where illumination is computed.

(b) Only diffuse illumination is evaluated for the hit points of the secondary rays traced in irradiance estimation. There are several good reasons for doing so:

- Irradiance caching relies on the smoothness of indirect illumination. However, specular highlights for the secondary rays may cause the indirect term to vary quite quickly, as exemplified by the appearance of caustics.

- The translation gradient calculation (Section 2.1.1.2) assumes that the incoming radiance samples $L_{j,k}$ do not change with translation. This assumption does not hold if specular reflection is taken into account at the secondary ray hit points, which results in incorrect gradient estimation. Irradiance interpolation with incorrect gradients usually leads to serious visual artifacts.

The ignored non-diffuse term for hit points of secondary rays corresponds to the caustics illumination. Ignoring caustics clearly underestimates the total indirect illumination at visible surfaces but it rarely leads to visually disturbing problems. Furthermore, caustics can be re-introduced by a specialized algorithm, such as photon mapping, as discussed in Section 4.3.3 below.

The computation of other terms in the above breakdown of the illumination integral is independent of irradiance caching and the algorithms used may vary from renderer to renderer. We refer to more general books, such as [PH04, DBB06], for more information. Nevertheless, let us briefly mention the most common approaches.

4.2.2   IDEAL SPECULAR REFLECTION

Ray tracing is used to resolve ideal specular reflection and transmission (Figure 4.3 b). Tracing a single reflected or transmitted ray, respectively, is sufficient. If a secondary ray generated by ideal specular reflection or transmission hit an area light source, the light source's self emission L_e is taken into account, in which way direct illumination is computed for the ideal specular term.

4.2.3   DIRECT ILLUMINATION

All renderers support direct illumination as it is often the most important illumination component. To compute direct illumination at a point, we iterate over the light sources and compute a contribution from each (Figure 4.3 d). Shadow maps [Wil78, SD02] or shadow rays [PH04] are used to generate shadows, i.e. to test visibility between the light source and the point being shaded. Illumination from area light sources is often resolved by sampling, effectively replacing the light source surface by a number of point lights [SWZ96]. Note that direct illumination computation does not take the ideal specular component of the BRDF into account since the probability density of a point on a light source being located exactly in the direction of specular reflection is zero.

4.2.4   GLOSSY INDIRECT ILLUMINATION

The glossy indirect illumination (or glossy reflection), is often the trickiest term to compute, since the BRDF properties that are classified under the label "glossy" can vary quite wildly: from sharp, mirror-like reflections, to dull, diffuse-like reflection. Glossy in our context simply means: neither ideal diffuse nor ideal specular.

4.2.4.1   BRDF Importance Sampling

For sharp, nearly specular reflection (see Figure 4.4, left), it is usually optimal to use distribution ray tracing optimized with BRDF-proportional importance sampling. Recall that importance sampling places more samples in directions where the BRDF has high value. Since the BRDF lobe for sharp reflections is very narrow, we can usually get away with only a few samples.

For dull, directional-diffuse reflection, (see Figure 4.4, right) the BRDF-proportional importance sampling is not very effective, since the BRDF lobe is quite wide. Nevertheless, distribution ray tracing is still often used to resolve this term, although a higher number of rays have to be used to suppress noise. However, similar to ideal diffuse surfaces, the indirect illumination on the dull glossy surfaces often changes slowly. We can take advantage of this and use illumination interpolation similar to irradiance caching, as described next.

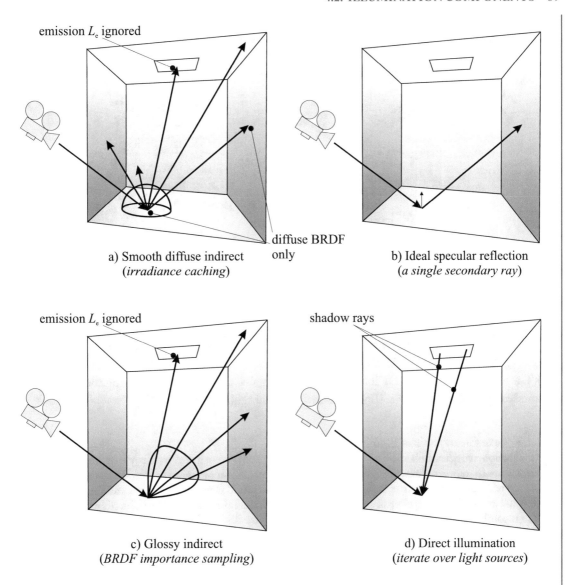

emission L_e ignored

diffuse BRDF only

a) Smooth diffuse indirect
(*irradiance caching*)

b) Ideal specular reflection
(*a single secondary ray*)

emission L_e ignored

shadow rays

c) Glossy indirect
(*BRDF importance sampling*)

d) Direct illumination
(*iterate over light sources*)

Figure 4.3: Specialized ray tracing techniques for the computation of various illumination components. Smooth indirect diffuse illumination is computed by irradiance caching (a). A single secondary ray is traced to solve specular reflections (b). Glossy reflections are usually solved using distribution ray tracing with BRDF-proportional importance sampling (c). Direct illumination is computed by light source sampling (d).

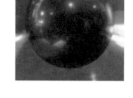

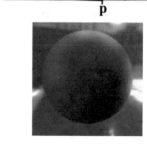

$f_{r\,(\mathbf{p},\omega_o)}(\omega_i)$ ω_o $f_{r\,(\mathbf{p},\omega_o)}(\omega_i)$ ω_o

p   **p**

(a) sharp glossy reflection   (b) dull glossy reflection

Figure 4.4: BRDF-proportional importance sampling works well for sharp glossy reflections, corresponding to narrow peaked BRDFs (a); less so for dull reflections corresponding to wide BRDF lobes (b). The top row shows the BRDF lobe shape for one outgoing direction ω_o while the bottom rows shows a corresponding rendered image. (Images courtesy of Addy Ngan.)

4.2.4.2   Fake Glossy Indirect Illumination

In this section, we describe Tabellion and Lamorlette's [TL04] extension of irradiance caching that allows to approximate indirect glossy illumination using the information gathered in irradiance calculation.

In irradiance caching, the only information that we retain from hemisphere sampling is the irradiance (plus its gradients). Effectively, we are discarding all the information about directional distribution of the incoming light. This is perfectly fine for the computation of diffuse illumination since it is view independent. However, surface appearance due to the glossy term depends on the relation of the viewing direction to the direction of incoming light. Therefore, if the cached information should be used for the computation of glossy illumination, we have to retain in the cache some information about the directionality of incoming light.

The approach of Tabellion and Lamorlette [TL04], described below, is extremely simple but often produces visually plausible results. Its motivation follows from the desire to leverage the existing infrastructure of their rendering system, where surface reflectance properties are represented by customizable pieces of code called *shaders*. On its input, a shader accepts the amount of incident light and the direction from which the light is arriving. On its output, it returns the surface color due to this light. Contributions from multiple lights is simply summed up.

The idea of Tabellion and Lamorlette is to convert the cached indirect illumination to a representation which, from the point of view of the shader inputs, is no different from a simple directional light source. The incoming radiance field is, therefore, approximated by the irradiance value along with *dominant directions* (one for each color component λ as shown in Figure 4.5). It is assumed that all the indirect light for a color component is coming fully from the corresponding dominant direction.

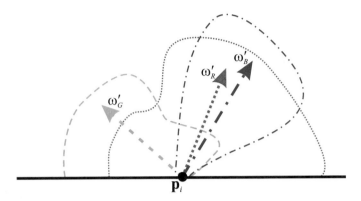

Figure 4.5: The directional distribution of incoming illumination at a point is approximated by three directional light sources, one for each color component. The light source directions correspond the dominant directions, ω_λ'.

A dominant directions ω_λ', $\lambda \in \{R, G, B\}$ is computed during hemisphere sampling, when a new cache record is created, as a normalized weighted average of the ray directions, with incoming radiance as the weight:

$$\omega_\lambda' = \frac{\sum_{j,k} \omega_{j,k} L_{j,k}^\lambda}{\| \sum_{j,k} \omega_{j,k} L_{j,k}^\lambda \|},$$

where $\omega_{j,k}$ is the sample direction and $L_{j,k}^\lambda$ is the corresponding radiance sample. (Refer to Section 2.1, pg. 19 for more details on indirect illumination sampling.) The dominant directions are stored in the cache along with the corresponding irradiance value.

When illumination is interpolated on a surface with a glossy component, the dominant directions are interpolated using a weighted average similar to that used for irradiance interpolation with the difference that the directions are re-normalized after interpolation:

$$\omega_\lambda'(\mathbf{p}) = \frac{\sum_{i \in S(\mathbf{p})} w_i(\mathbf{p}) \, \omega_{\lambda,i}'}{\| \sum_{i \in S(\mathbf{p})} w_i(\mathbf{p}) \, \omega_{\lambda,i}' \|},$$

where $S(\mathbf{p})$ is the set of irradiance records contributing to interpolation at $\mathbf{p}$ and $w_i(\mathbf{p})$ is the interpolation weight of record i at point $\mathbf{p}$. (Refer to Section 2.2.1, pg. 27 for more details on interpolation.)

Since surface shaders are usually designed to accept *radiance* as the quantity describing the amount of incident light, the interpolated irradiance value $E(\mathbf{p})$ is converted to a radiance value. Assuming a purely diffuse surface, the conversion is as follows:

$$L_\lambda(\mathbf{p}, \omega_\lambda') = \frac{E_\lambda(\mathbf{p})}{|\omega_\lambda' \cdot \mathbf{n}|}.$$

Now, $L_\lambda(\mathbf{p}, \omega'_\lambda)$ along with the corresponding dominant direction ω'_λ can be though of as one directional light source. Finally, the surface color due to glossy indirect illumination is computed by invoking the surface shader with the "indirect lights" on the input.

The above techniques assumes that light is coming from a single dominant direction, which is often not true. Still, the technique works fairly well for low-gloss surfaces with an irregular structure, where the human visual system does not easily recognize the inaccuracy (Figure 4.6 left). For smooth surfaces, the approximation is often obvious and even disturbing (Figure 4.6 right).

(a) Bumped surface   (b) Smooth surface

Figure 4.6: The fake glossy indirect illumination works fine for surfaces with irregular structure (a), since the human visual system is unable to tell the approximation. On smooth surfaces, though, the approximation may be obvious.

4.2.4.3   Radiance Caching

Radiance caching, proposed by Křivánek et al. [KGPB05], is a more accurate algorithm for computing indirect illumination on glossy surfaces. Similar to irradiance caching, it is based on sparse sampling and interpolation of illumination. Since radiance caching interpolates on glossy surfaces, the directional distribution of indirect lighting has to be stored in the cache. The representation by spherical harmonics used in radiance caching for that purpose is more accurate than a simple approximation with directional lights used by Tabellion and Lamorlette. We refer to the original paper or to [Kri05] for more details.

4.3   RECURSION

Hemisphere sampling in irradiance caching and in the calculation of glossy and specular components relies on obtaining samples of indirect radiance L_i. These samples are computed by tracing secondary rays. But what calculation takes place at the hit points of these secondary rays?

The radiance samples are calculated by evaluating the illumination integral recursively for the hit points of the secondary rays:

$$L_i(\mathbf{p}, \omega_i) = L_o(\text{isect}(\mathbf{p}, \omega_i), -\omega_i),$$

where the function isect($\mathbf{p}$, ω_i) finds the nearest intersection of a ray from $\mathbf{p}$ in the direction ω_i (see Figure 4.7).

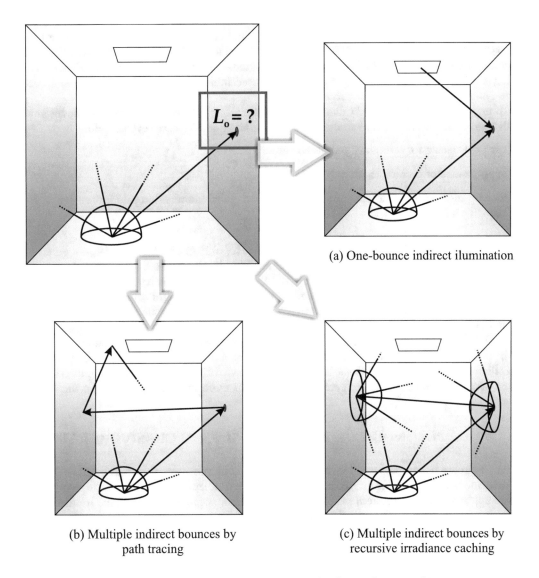

(a) One-bounce indirect ilumination

(b) Multiple indirect bounces by path tracing

(c) Multiple indirect bounces by recursive irradiance caching

Figure 4.7: Different approaches to compute radiance samples for irradiance caching.

No matter what technique we use for the illumination integral evaluation at the secondary hit points, we must respect a general rule of thumb: *The self emission L_e is ignored at the hit points of the secondary rays generated when sampling the indirect glossy and diffuse terms (including sampling in irradiance caching).* The reason is that the hemisphere sampling is only used to compute indirect illumination. Direct illumination from the light sources is evaluated by a different technique. However, when sampling ideal specular reflection and transmission, the self emission L_e is *not ignored*.

A simple option for evaluating the illumination integral for the secondary hit points is to consider the *direct illumination only*, as shown in Figure 4.7(a). This yields an incomplete global illumination solution commonly referred to a "*one-bounce indirect*" illumination. Tabellion and Lamorlette [TL04] points out that in the context of cinematic lighting, the one-bounce indirect illumination is sufficient since it provides the desired smooth shading and shadowing gradations; the actual lighting levels are tweaked manually anyway. Obviously, for predictive rendering used in illumination engineering or architecture, the one-bounce approximation is not acceptable.

If multiple bounces of indirect illumination are desirable, the radiance estimate at the hit point of secondary rays must itself take into account indirect illumination. We have several options here:

1. use path tracing for secondary and higher-order bounces,

2. apply irradiance caching recursively,

3. look up indirect illumination from a pre-computed global illumination solution, such as a *photon map*.

4.3.1   MULTIPLE INDIRECT BOUNCES USING PATH TRACING

The simplest way to include multiple bounces of indirect illumination is to use path tracing for secondary and higher-order bounces as shown in Figure 4.7 (b). In path-tracing, a single random ray is traced to estimate the indirect illumination. Direct illumination uses the classic light source sampling algorithm. For unbiased results, the recursion in path tracing may be terminated by using a technique known as Russian roulette [DBB06].

Path tracing is general and flexible but its main disadvantage is noise: each radiance sample computed by path tracing is noisy, therefore, we need more rays in hemisphere sampling to obtain a reliable indirect illumination estimate. This issue is critical for irradiance caching since a) the computed irradiance is reused over a large area and b) the gradient estimation is more sensitive to noise than the irradiance estimate itself.

4.3.2   MULTIPLE INDIRECT BOUNCES USING RECURSIVE IRRADIANCE CACHING

Another possibility to compute multiple bounces of indirect illumination, shown in Figure 4.7 (c), is to apply irradiance caching recursively. Recursive application of irradiance caching is used for example in the *Radiance* lighting simulation system [WLS98]. It is important not to mix together the irradiance values computed at different recursion levels. This can be achieved for example by keeping a separate irradiance cache for each recursion level.

The error of irradiance calculation at higher recursion levels has relatively low impact on the final image error. We can take advantage of this and use fewer rays for hemisphere sampling at higher recursion levels and allow more error in interpolation by increasing the a value. Assuming 50% average surface reflectivity, halving the number of rays at each recursion level and multiplying the allowed error a by $\sqrt{2}$, the image error contributed from each level of recursion will be the same [WRC88]. In practice, due to perception effects (people are much more sensitive to "directly visible" errors), we can reduce the number of rays for higher recursion levels even more drastically.

Figure 4.8 illustrates the behavior of recursive irradiance caching. First, the cache is empty for all recursion levels. As soon as the first ray hits a surface, the recursion propagates to the higher levels. The irradiance cache is first filled for the highest level, then gradually for lower levers etc. Towards the end of the image rendering, the recursion rarely exceeds level two since most queries can be satisfied by the cache contents at the low recursion levels.

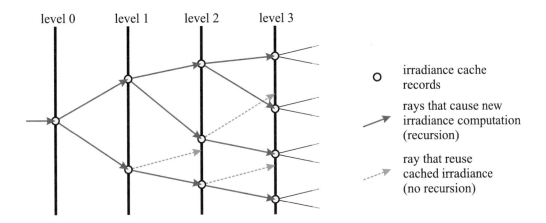

Figure 4.8: Filling of caches in recursive irradiance caching causes the recursion to stop at lower and lower levels as the computation progresses, since the rays that reuse cached irradiance do not spawn any further rays. (Redrawn after [WRC88].)

Compared to path tracing, recursive irradiance caching has the advantage that indirect illumination on the higher levels is "smoothed out" by interpolation, i.e. noise-free (although biased), and, therefore, we can get away with fewer rays for hemisphere sampling on the first level. In addition, once we have a working implementation of irradiance caching, extension to a recursive version is simple.

4.3.3 MULTIPLE INDIRECT BOUNCES USING PHOTON MAPPING

Photon mapping [Jen01] is a two-pass technique for global illumination computation that consists of 1) the photon tracing pass and 2) the image rendering pass (see Figure 4.9). Irradiance caching is used in the image rendering pass.

4.3.3.1 First pass: Photon tracing

In the photon tracing pass, particles called photons are emitted from the light sources and traced through the scene in a similar way that path tracing follows paths from the camera. Every time a photon hits a diffuse (or dull glossy) surface, the hit position along with the photon energy is recorded in a data structure called the *photon map*. At the end of the photon tracing pass, the photon map contains a representation of global illumination in the scene.

As a matter of fact, two photon tracing passes are carried out. The output of each is stored in a separate photon map (see Figure 4.10):

Caustics photon map. Contains only caustics illumination, that is the light that is reflected or refracted only on specular/sharp glossy surfaces before eventually arriving on a diffuse surface (i.e. LS^+D paths). Since the caustics patterns are usually concentrated due to light focusing by specular surfaces,

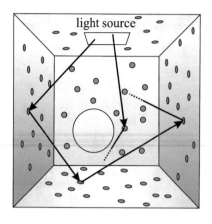

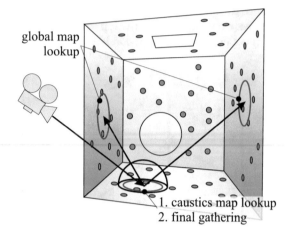

Pass 1: Photon tracing   Pass 2: Image rendering

Figure 4.9: Global illumination computation with photon mapping consists of two passes: 1) the photon tracing pass (left) and 2) the image rendering pass (right). (Redrawn after [Dut03].)

an accurate representation of caustics can be obtained with a reasonable number of photons ($\approx$ 500,000).

Global photon map. Contains the representation of the overall illumination in the entire scene, including caustics, direct, and diffuse indirect illumination, i.e. the $L(S|D)^+D$ paths. The global map contains a rather rough information since too many photons would be required for high accuracy.

4.3.3.2   Radiance Estimate and Photon Visualization

The fundamental operation that is used in rendering with photon maps is the *radiance estimate*: given a point **p** and outgoing direction ω_o, a look up in a photon map gives us an estimate of reflected radiance $L_r(\mathbf{p}, \omega_o)$ without tracing any rays. The radiance estimate is based on density estimation using N_p (usually 50-250) nearest photons [Jen01]:

$$L_r(\mathbf{p}, \omega_o) = \frac{1}{\pi r^2} \sum_{p=1}^{N_p} f_r(\mathbf{p}, \omega_{i,p}, \omega_o) \Delta \Phi_p,$$

where

$\omega_{i,p}$ is the incident direction of the p-th nearest photon,
$\Delta \Phi_p$ is the flux carried by the p-th nearest photon, and
r is the distance to the farthest of the N_p nearest photons (i.e. the search radius).

For our purposes, it will be sufficient to treat the radiance estimate as a black box.

The caustics map gives an accurate estimate of caustics illumination. This is why it can be directly visualized, i.e. we can use a caustics map query for the primary ray hit points (see Figure 4.9 right).

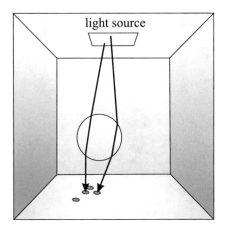
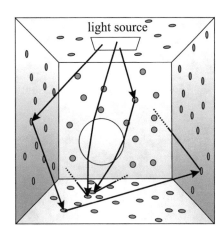

Caustics photon map   Global photon map

Figure 4.10: Two separate photon maps are created in the photon tracing pass: The causics map (left) represents caustics illumination only (LS^+D paths). The global map (right) is a rough representation of complete global illumination ($L(S|D)^+D$ paths). (Redrawn after [Dut03].)

The information stored in the global photon map, on the other hand, is less accurate. Using a global photon map estimate directly for primary rays generates splotches in the images, as shown in Figure 4.11 (a). This is why radiance estimates from the global photon map are only used after one step of distribution ray tracing (Figure 4.11 (b)). This distribution ray tracing step is often referred to as *final gathering* and is usually the bottleneck of the whole rendering process. Fortunately, irradiance caching can significantly speed up the final gathering. Let us now summarize the image rendering step of the photon mapping algorithm.

4.3.3.3   Second pass: Image rendering

When rendering with photon mapping, we distinguish the "accurate" and the "approximate" calculation of each illumination component. The accurate calculation is used for directly visible points, or points after few specular reflections, or after non-specular reflection when the ray length is below a threshold (in order to avoid errors due to photon map inaccuracy in corners). The approximate calculation is applied at the locations accessed after a diffuse reflection.

Direct illumination uses light source sampling for accurate evaluation and a global map radiance estimate for the approximate evaluation. *Ideal specular reflection* uses ray tracing (both for the accurate and approximate evaluation). *Caustics* are resolved by a lookup in the caustics map for the accurate evaluation. For the approximate evaluation, the global map radiance estimate already includes the caustics.

The smooth indirect illumination is solved by the final gathering discussed above, accelerated by irradiance caching. The approximate evaluation of the smooth indirect illumination is, again, included in the global photon map estimate.

The accurate evaluation of the *glossy indirect term* may use distribution ray tracing with BRDF-proportional importance sampling, or alternatively some of the caching approaches described in Section 4.2.4.2 earlier in this chapter. The global map radiance estimate may be used for the approximate

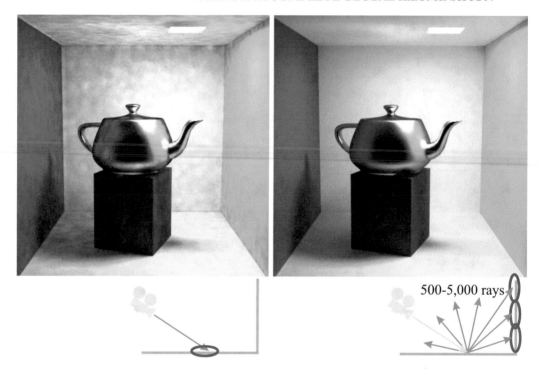

(a) global map used for primary rays   (b) final gathering with irradiance caching

Figure 4.11: Global photon map visualization. Since the representation of global illumination in the photon map is quite rough, visualizing the global map directly for primary rays produces splotchy images (a). This is why one step of distribution ray tracing (called *final gathering*) is used before the global map lookups (b). The final gathering can be greatly accelerated by the use of irradiance caching.

evaluation but only for rather diffuse surfaces. For sharp glossy reflections, it is better to use distribution ray tracing with BRDF-proportional sampling. (For the approximate evaluation we would usually trace only a single random ray.)

To summarize, a single radiance estimate in the global photon map gives direct illumination, glossy and diffuse indirect illumination, and caustics. Unfortunately, global maps contains only a rough information, which is why it is used only for the approximate evaluation, whereas the accurate evaluation applies a specialized technique for each component.

4.3.3.4   Faster Radiance Estimates
A photon map represents illumination as a collection of a large number of photons (in the order of millions). The radiance estimate at a point is based on locating N_p (50-250) nearest photons. This nearest neighbor query may be rather costly, which is why alternative methods that allow faster radiance estimates were developed (for the global photon map). Usually, they only work for Lambertian surfaces. For example, Christensen et al. [CB04, Chr05] use so-called *brick maps*. Tabellion and Lamorlette [TL04] use radiosity textures.

As an optimization, Christensen [Chr00] pre-computes the radiance estimate at photon locations in a pre-process. To perform a radiance estimate at a point during image rendering, he simply locates a single nearest photon and retrieves its pre-computed radiance estimate. However, the disadvantage of this method is its view independence: the radiance estimates are pre-computed in the whole scene, even in locations where none may be needed.

Using the lazy-evaluation scheme of irradiance caching, we can improve on Christensen's technique. We associate an irradiance cache with the global photon map. To perform a radiance estimate at a point, we first query the cache. If nothing is found, then we perform the actual photon map radiance estimate and store the result in the cache. We set the "distance to surfaces" R_i equal to r (i.e. the search radius in the photon map radiance estimate). We use the approximation error $a = 0.5$ so that the validity radius of a record, $a R_i$, is half the value of r. This technique reduces the number of costly photon map radiance estimates and performs them only where needed.

4.3.3.5   Discussion

The combination of photon mapping and irradiance caching results is an efficient method for complete global illumination computation, since the advantages of the two algorithms are complementary. Photon mapping requires final gathering since direct visualization of the global map is not accurate enough. Without irradiance caching, the final gathering is slow. On the other hand, the fact that photon mapping separates the caustics illumination makes the indirect diffuse lighting truly smooth and consequently, the irradiance interpolation in irradiance caching can reliably produce high-quality images (see Figure 4.12).

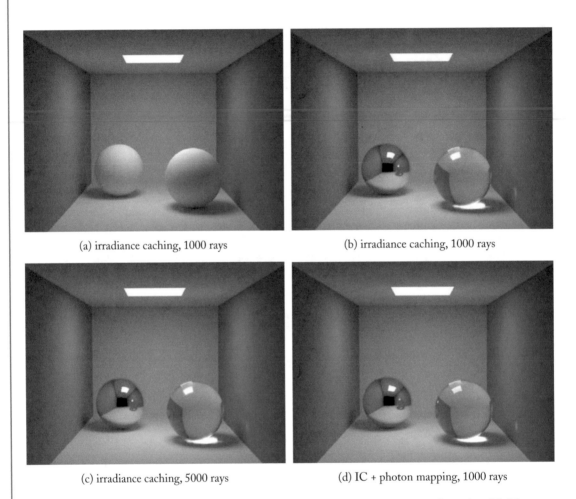

(a) irradiance caching, 1000 rays   (b) irradiance caching, 1000 rays

(c) irradiance caching, 5000 rays   (d) IC + photon mapping, 1000 rays

Figure 4.12: (a) In a purely diffuse scene, irradiance caching alone gives good results. (b) However, making some objects specular and refractive creates caustics, which break the underlying assumption of irradiance caching that indirect illumination changes slowly over surfaces. As a consequence irradiance caching produces visible low-frequency noise. (c) Increasing the number of rays per hemisphere reduces most of the low-frequency noise but the caustics are still not properly resolved due to interpolation. (d) The combination of irradiance caching with photon mapping produces good results: irradiance caching calculates the smooth illumination while photon mapping adds the highly varying caustics. (Images courtesy of Henrik Wann Jensen.)

CHAPTER 5

Irradiance Caching on Graphics Hardware

In the classical irradiance caching algorithm, the irradiance cache is queried for each visible point in the scene to determine if a new irradiance record should be created. When a new record is required, the irradiance value is computed by sampling the surrounding hemisphere using ray tracing. The newly created record is then stored in an octree, called the irradiance cache. For points nearby the record location, the record can later be retrieved by querying the octree.

In this chapter, we consider the use of graphics hardware for fast global illumination computation using irradiance caching. However, besides the high computational power of graphics processors, efficiency is highly dependent on the compatibility of the implemented algorithm with the computational model of the GPU (graphics processing unit). More specifically, stream processors, such as GPUs, are massively parallel, and hence are not very efficient for dealing with the repeated branching inherent to recursive structures such as octrees. Also, as pointers are not natively supported, the implementation of an unbalanced octree is not straightforward. To achieve a better efficiency, we reformulate the irradiance caching algorithm to better fit to the computational model of graphics hardware: parallel computing, simple data structures, and limited conditional statements.

To this end, we first revisit the algorithm for irradiance interpolation within the validity areas of irradiance records. We propose the *irradiance splatting* as its efficient version targeted to the GPU. Furthermore, as GPUs do not natively support ray tracing and traversal of acceleration structures, we also propose *GPU-based hemisphere sampling* for the computation of one-bounce indirect illumination.

Algorithm 13 Irradiance Caching on Graphics Hardware

Render the scene and store information about visible points:
location $\mathbf{p}$, surface normal $\mathbf{n}$, material reflectance ρ_d
Splat the contents of the irradiance cache
for all pixels **do**
    if Pixel has no irradiance contribution yet **then**
        Sample the hemisphere above point $\mathbf{p}$ visible through pixel
        Create a new cache record
        Splat the record contribution onto the irradiance splat buffer
    end if
end for
Render the records onto the image plane

The GPU implementation is organized as described in Algorithm 13. The scene is first rendered from the viewpoint to obtain information on the visible points: their location $\mathbf{p}$, surface normal $\mathbf{n}$ and material reflectance ρ_d. Then, the possible existing contents of the irradiance cache are *splatted* onto an *irradiance splat buffer* (Section 5.1). If some new records are needed to render the image, the GPU is used to sample the hemisphere (Section 5.3) and the new records also get stored in the cache and splatted. The cache is finally rendered on the GPU using the irradiance splat buffer (Section 5.4).

5.1 IRRADIANCE SPLATTING

The irradiance splatting algorithm approaches the irradiance interpolation problem backwards compared to the octree-based method: instead of starting from a point and looking for the nearby records, this method starts from a record and accumulates its contribution to the lighting at each visible point of the scene by splatting the record onto a *splat buffer*, which has the same size as the frame buffer. Each pixel SplatBuff(x, y) of the splat buffer is a pair (E, w), where E is the sum of the weighted contributions of the records, and w is the cumulated weight of those contributions.

The irradiance splatting (Algorithm 14) is designed for computing the contribution of an irradiance record to the indirect lighting of visible points. The derivation starts from the equation used for irradiance interpolation in the classic irradiance caching:

$$E(\mathbf{p}) = \frac{\sum_{i \in S(\mathbf{p})} E_i(\mathbf{p}) w_i(\mathbf{p})}{\sum_{i \in S(\mathbf{p})} w_i(\mathbf{p})} \tag{5.1}$$

The indirect irradiance $E(\mathbf{p})$ at point $\mathbf{p}$ is estimated as the weighted average of the contributions of nearby irradiance records evaluated at point $\mathbf{p}$. The weight allocated to a record i at point $\mathbf{p}$ with normal $\mathbf{n}$ is defined in Section 2.2.1, pg. 27, as

$$w_i(\mathbf{p}) = \frac{1}{\frac{\|\mathbf{p}-\mathbf{p}_i\|}{R_i} + \sqrt{1 - \mathbf{n} \cdot \mathbf{n}_i}} - \frac{1}{a} \tag{5.2}$$

where $\mathbf{p}_i$, $\mathbf{n}_i$ and R_i are respectively the location of record i, its normal, and the distance to surfaces visible from $\mathbf{p}_i$ (see Section 2.2.2, pg. 29). A user-defined value a represents the accuracy of the interpolation. This value is used to threshold the records' validity area: record i contributes to irradiance interpolation at point $\mathbf{p}$ if and only if

$$w_i(\mathbf{p}) > 0 \tag{5.3}$$

Substituting Equation (5.2) into Equation (5.3) and assuming a flat surface, *i.e.* $\mathbf{n} = \mathbf{n}_i$, one can see that a necessary condition for record i to contribute to irradiance interpolation at point $\mathbf{p}$ is:

$$\|\mathbf{p} - \mathbf{p}_i\| \leq a R_i \tag{5.4}$$

Therefore, Equation (5.4) guarantees that a record i cannot contribute to interpolation at any point outside a sphere I_i centered at $\mathbf{p}_i$, with radius $r_i = a R_i$.

Given the camera parameters, the *irradiance splatting* splats the sphere I_i onto the image plane (Figure 5.1). In practice, this splatting is implemented in a vertex shader (see Algorithm 15), where the record properties are stored in the attributes of each vertex of a quadrilateral $(0, 0)$, $(1, 0)$, $(1, 1)$, $(0, 1)$. Note that, while using the point primitive instead of a quad would be more efficient, graphics hardware have a limited maximum point size, which may complicate the splatting of records with a large validity area. Alternatively, geometry shaders can be used to generate the splats on a Shader Model 4-compliant graphics hardware.

Algorithm 14 Irradiance splatting

Let $i = \{\mathbf{p}_i, \mathbf{n}_i, E_i, R_i\}$ be the record being splatted
Let I_i be the sphere centered at $\mathbf{p}_i$ with radius $a R_i$ (a is the accuracy parameter)

Determine the bounding box of sphere I_i on the image plane
for all pixels $P(x, y) = \{\mathbf{p}, \mathbf{n}, \rho_d\}$ in the bounding box **do**
    // Evaluate weight at $\mathbf{p}$
    $w_i(\mathbf{p}) \leftarrow \dfrac{1}{\frac{\|\mathbf{p}-\mathbf{p}_i\|}{R_i}+\sqrt{1-\mathbf{n}\cdot\mathbf{n}_i}} - \dfrac{1}{a}$
    if $(w_i(\mathbf{p}) > 0)$ **and** $(\mathbf{p}_i$ not in front of $\mathbf{p})$ **then**
        // Extrapolate irradiance of record i at point $\mathbf{p}$
        $E_i(\mathbf{p}) \leftarrow E_i + (\mathbf{n}_i \times \mathbf{n}) \cdot \nabla_r E_i + (\mathbf{p} - \mathbf{p}_i) \cdot \nabla_t E_i$
        // Accumulate into the irradiance splat buffer
        SplatBuff$(x, y).E$ += $w_i(\mathbf{p})E_i(\mathbf{p})$
        SplatBuff$(x, y).w$ += $w_i(\mathbf{p})$
    end if
end for

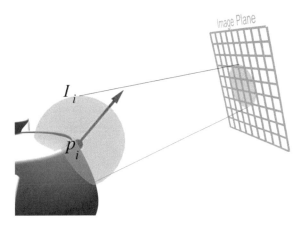

Figure 5.1: The sphere I_i around the position $\mathbf{p}_i$ of the record i is splatted onto the image plane. For each point within the sphere splat, the contribution of record i is accumulated into the irradiance splat buffer.

The vertex attributes contain some of the fields of the record: position $\mathbf{p}_i$, surface normal $\mathbf{n}_i$, distance to surfaces R_i, irradiance E_i and its gradients $\nabla_r E_i$ and $\nabla_t E_i$. For the sake of efficiency, the inverse squared distance to surfaces, i.e. $1/R_i^2$, is also passed in a vertex attribute.

Algorithm 15 Pseudo-code for irradiance splatting: Vertex Shader

---

```
uniform float a;

attribute vec4 recordPosition;
attribute vec4 recordNormal;
attribute vec3 irradiance;
attribute vec3 rotGradR, rotGradG, rotGradB;
attribute vec3 transGradR, transGradG, transGradB;
attribute float Ri;
attribute float sqInvRi;

// Compute record position in camera space
vec4 camSpacePosition = ModelViewMatrix * recordPosition;
// Compute the splatted validity sphere in camera space
vec4 splatSphereBox(a*Ri, a*Ri, 0, 0);
// Compute the actual corners of the box in screen space
vec4 topRight = ProjectionMatrix * (camSpacePosition + splatSphereBox);
topRight /= topRight.w;
vec4 bottomLeft = ProjectionMatrix*(camSpacePosition - splatSphereBox);
bottomLeft /= bottomLeft.w;
// Infer the projected vertex position
vec2 delta = topRight.xy - bottomLeft.xy;
outputVertex = vec4(
      bottomLeft.x + inputVertex.x*delta.x,
      bottomLeft.y + inputVertex.y*delta.y,
      0.0
      1.0);
```
Directly pass the vertex attributes to the fragment shader

The weighting function given by Equation (5.2) is evaluated for each point visible through pixels covered by I_i. For each pixel with positive weight, the algorithm computes the contribution of record i to the irradiance estimate (Algorithm 16).

Note that this splatting approach can be used with any weighting function containing a distance criterion. In this chapter we focused on Ward et al.'s weighting function [WRC88], although the weight proposed in [TL04] as discussed in Section 2.2.1 could be employed as well.

5.2   CACHE MISS DETECTION

The rendering process of irradiance caching requires the generation of a number of irradiance records to obtain a reasonable estimate of the indirect lighting for each point visible from the camera. Therefore,

---

Algorithm 16 Pseudo-code for irradiance splatting: Fragment Shader

---

```
// Textures containing hit positions and hit normals
uniform sampler2D hitPosTex, hitNormTex;
// Target accuracy
uniform float a;
// Squared target accuracy, a²
uniform float squaredA;


// Determine the texture coordinates for fetching scene information
vec2 textureCoord = vec2(pixelCoord.x,pixelCoord.y);
// Retrieve the corresponding hit position and surface normal
vec3 hitPos    = texture2D(hitPosTex,  textureCoord).xyz;
vec3 hitNormal = texture2D(hitNormTex, textureCoord).xyz;
// Compute the contribution weight
float weight = computeWeight();
// Discard the fragment if weight is negative
if ( weight <= 0 || inFrontTest()==true ) kill;
// The record is usable for interpolation at the pixel position – compute contribution
 // a) Compute the translation gradient component
vec3 diff = hitPos - recordPosition;
vec3 transGradContrib = vec3(
      dot(diff, transGradR),
      dot(diff, transGradG),
      dot(diff, transGradB) );
 // b) Compute the rotation gradient component
vec3 normalsCrossProduct = cross(recordNormal, hitNormal);
vec3 rotGradContrib = vec3(
      dot(normalsCrossProduct, rotGradR),
      dot(normalsCrossProduct, rotGradG),
      dot(normalsCrossProduct, rotGradB) );
FragmentColor.rgb = weight * (
      irradiance +
      transGradContrib +
      rotGradContrib );
FragmentColor.a = weight;
```

---

even though the values of the records are view-independent, their locations are determined according to the current viewpoint.

Usually, the irradiance cache records are stored in an octree for fast lookups. For a given pixel of the image, a ray is traced in the scene to find the point **p** visible through this pixel. At this point, the octree is queried to determine whether a new record is required. If yes, the actual indirect lighting is computed at **p**: a record is created and stored in the cache. Otherwise the cache is used to compute an estimate of the indirect lighting using interpolation. Since each visible point is processed only once, a newly created record cannot contribute to the indirect lighting of points for which the indirect lighting has already been computed or estimated. Therefore, unless an appropriate image traversal algorithm is employed, artifacts may be visible (Figure 5.2(a)). The traversal typically relies on hierarchical subdivision of the image to reduce and spread out the errors (see Section 3.1, pg. 43). Artifact-free images can only be obtained in two passes: in the first pass, the records required for the current viewpoint are generated. In the second pass, the scene is rendered using the contents of the cache.

On the other hand, the irradiance splatting algorithm is based on the independent splatting of the records. The contribution of a record is splatted onto each visible point within its validity area, even though the points have been previously checked for cache miss (Figure 5.2(b)). Therefore, this method avoids the need of a particular image traversal algorithm without harming the rendering quality. For the sake of efficiency and memory coherence, a linear (scanline) traversal of the image is recommended. Note that any other image traversal algorithm could be used.

Once the algorithm has determined where a new record has to be generated, the irradiance value and gradients must be computed by sampling the hemisphere above the new record position. The following section details a simple method for record generation using graphics hardware in the context of one-bounce global illumination computation.

5.3   GPU-BASED RECORD GENERATION

In the method described hereafter, the incoming irradiance and gradients associated with a record are generated using both the GPU and CPU. First, the GPU performs the actual hemisphere sampling. Then, the CPU computes the corresponding irradiance and gradients.

5.3.1   HEMISPHERE SAMPLING ON THE GPU

In the context of ray tracing-based global illumination, the visibility tests are generally the bottleneck of the algorithms. However, the hemisphere sampling for irradiance estimation can be sped up by using the GPU rasterization instead of ray tracing. In the classical hemi-cube method [CG85], five rendering passes are necessary for full hemisphere sampling on graphics hardware. For the sake of efficiency, our algorithm uses a one-pass method similar to [SP89, LC04]: using a virtual camera placed at the record location, the rasterization engine samples the hemisphere above the record to compute the radiances incoming from each direction (Figure 5.3).

A typical field of view of the virtual camera used for hemisphere sampling is twice the aperture of the standard hemicube, that is approximately 126.87 degrees. More specifically, the side length of the image plane ($imgSize$) is four times the distance of the image plane from the center of projection ($near$), hence the field-of-view angle:

$$fov = 2 \arctan\left(\frac{imgSize/2}{near}\right) = 2 \arctan(2) \approx 126.87 \text{ deg.}$$

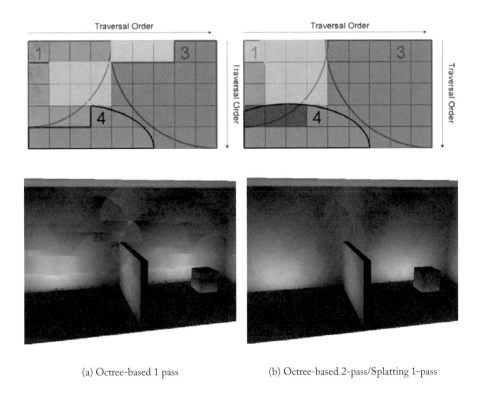

(a) Octree-based 1 pass   (b) Octree-based 2-pass/Splatting 1-pass

Figure 5.2: Upper row: the irradiance records 1, 2, 3, 4 and their respective validity areas. We consider the use of a scanline traversal of the image. In the classical, octree-based method, a newly created record cannot contribute to pixels which have already been examined. Therefore, discontinuities appear in the validity area of the records, yielding artifacts. Those artifacts are illustrated in the upper row: record 1 is created at the point visible through the first pixel, and used for estimating the indirect lighting in the next 3 pixels. When record 2 is created at the point visible through the fifth pixel, its contribution is not propagated to the previous pixels, hence creating a discontinuity. A second rendering pass is needed to generate an image without such discontinuities. Using the irradiance splatting method, each record contributes to all possible pixels, avoiding the need of a second rendering pass. Lower row: the indirect lighting obtained using these methods.

However, as shown in Figure 5.4, using a single rasterization plane cannot account for the radiances coming from grazing angles. To compensate for the incomplete hemisphere coverage, Larsen *et al.* [LC04] divide the obtained irradiance by the plane coverage ratio. A more accurate method consists in virtually stretching border pixels to fill the remaining solid angle (Figure 5.4). This method generally yields more accurate results than the approach of Larsen *et al.* (Table 5.1). Furthermore, a key aspect of this method is

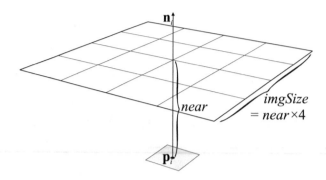

Figure 5.3: Hemisphere sampling is implemented on the GPU by placing a camera at the record location and rendering the scene.

that the directional information of the incoming radiance remains plausible. Therefore, unlike in [LC04], the indirect glossy lighting could also be rendered correctly if radiance caching [KGPB05] is used.

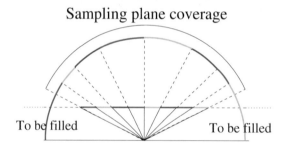

Figure 5.4: Hemisphere sampling is reduced to the rasterization of the scene geometry onto a single sampling plane. Since this plane does not cover the whole hemisphere, we use a border compensation method to account for the missing directions. Border pixels are virtually stretched to avoid zero lighting coming from grazing angles, yielding more plausible results.

GPU-based hemisphere sampling requires shading of the points visible by the virtual camera, accounting for direct lighting and shadowing. This computation is also carried out by the graphics hardware using per-pixel shading and shadow mapping. We recommend using uniform shadow maps [Wil78]: unlike perspective shadow maps [SD02] or shadow volumes [Hei91], uniform shadow maps are view-independent: in static scenes the shadow map is computed once and reused for rendering the scene from each viewpoint. Therefore, the algorithm reuses the same shadow map to compute all the records, hence reducing the shading cost. Furthermore, the incoming radiance values undergo averaging to compute irradiance. This averaging blurs out aliasing artifacts of the shadow maps, hence providing inexpensive, high quality irradiance values.

Once the hemisphere has been sampled, the algorithm computes the actual irradiance record.

Table 5.1: RMS error of 10000 irradiance values computed in the Sibenik Cathedral scene (Figure 5.9). The border stretching method yields more accurate results than other approaches, while preserving the directional information of the incoming radiance. The reference values are computed by full hemisphere sampling using Monte Carlo integration.

	Plane sampling [SP89]	Coverage ratio [LC04]	Border stretching [GKBP05]
RMS Error	18.1%	10.4%	5.8%

5.3.2   IRRADIANCE COMPUTATION

Irradiance is defined as a hemispherical integral of incoming radiance weighted by the cosine term (see Equation (2.1) on pg. 18). We evaluate the integral numerically as a sum of incoming radiance values of the pixels on the sampling plane, $L_{j,k}$, weighted by each pixel's solid angle, $\Omega_{j,k}$, and the corresponding cosine term, $\cos\theta_{j,k}$:

$$E \approx \sum_{k=0}^{N-1} \sum_{j=0}^{N-1} L_{j,k}\, \Omega_{j,k} \cos\theta_{j,k},$$

where $N \times N$ is the resolution of the sampling plane in pixels. The $\Omega_{j,k} \cos\theta_{j,k}$ term in the sum can be pre-computed to accelerate irradiance computation.

The solid angle of a pixel is approximated using an expression for the solid angle of a differential area, i.e. $d\omega = dA \cos\theta / d^2$. For a camera with the field-of-view angle given by fov, the solid angle of a pixel amounts to:

$$\Omega_{j,k} = \underbrace{\left(\frac{2\,near \tan\frac{fov}{2}}{N}\right)^2}_{\text{pixel area}} \frac{\cos\theta_{j,k}}{d_{j,k}^2},$$

where $near$ is the distance of the image plane from the center of projection and $d_{j,k}$ is the distance of the pixel center form the center of projection. In our implementation, the field-of-view of 126.87 degrees yields the pixel area of $(4\,near/N)^2$, hence the pixel solid angle: $\Omega_{j,k} = \left(\frac{4\,near}{N}\right)^2 \frac{\cos\theta_{j,k}}{d_{j,k}^2}$.

As explained in the previous section, this sampling technique leaves a part of the hemisphere unsampled. To compensate for the lack of information, the pixels located on the edge of the sampling plane are virtually extended. Using the notation of Figure 5.6, the solid angle covered by an edge pixel (j, k) is approximately:

$$\Omega_{j,k}^{edge} = \Omega_{j,k} + (1 - \cos\theta_{max})(\phi_{max} - \phi_{min}) \tag{5.5}$$

For the pixel highlighted in Figure 5.6, the angles above are derived from the vector $\mathbf{d}$ from the sampling point to the center of the pixel. To compute θ_{max}, let us consider the vector $\mathbf{d}^{\theta_{max}}$ from the sampling point to the edge of the sampling plane:

$$\mathbf{d}^{\theta_{max}} = \mathbf{d} + (\delta_x, 0, 0). \tag{5.6}$$

The angle θ_{max} can then be obtained as:

$$\theta_{max} = \arccos \frac{\mathbf{d}_z^{\theta_{max}}}{\|\mathbf{d}^{\theta_{max}}\|}. \tag{5.7}$$

Following the same principle, we obtain ϕ_{min} and ϕ_{max}:

$$\mathbf{d}^{\phi_{min}} = \mathbf{d} + (\delta_x, -\delta_y, 0) \tag{5.8}$$
$$\mathbf{d}^{\phi_{max}} = \mathbf{d} + (\delta_x, +\delta_y, 0). \tag{5.9}$$

The angles are then:

$$\phi_{min} = \arctan \mathbf{d}_y^{\phi_{min}}/\mathbf{d}_x^{\phi_{min}} \tag{5.10}$$
$$\phi_{max} = \arctan \mathbf{d}_y^{\phi_{max}}/\mathbf{d}_x^{\phi_{max}}. \tag{5.11}$$

For the purpose of the gradient computation, we define the direction corresponding to an edge pixel as:

$$(\theta_{j,k}^{edge}, \phi_{j,k}^{edge}) = (\frac{1}{2}(\theta_{j,k} + \frac{\pi}{2}), \phi_{j,k}). \tag{5.12}$$

By using $\Omega_{j,k}^{edge}$ and $(\theta_{j,k}^{edge}, \phi_{j,k}^{edge})$ for edge pixels instead of the regular values, the border extension method gets seamlessly integrated into the irradiance and gradient estimation process.

Implementation   The irradiance sum can be calculated either on the GPU using automatic mipmap generation [LC04], or using frame buffer readback and CPU-based summation (Figure 5.5). In the case of irradiance caching, a record contains the irradiance, the rotational and translation gradients, and the harmonic mean distance to the objects visible from the record location. These values can be stored as 22 floating-point numbers. Hence, the GPU-based summation of [LC04] would require the computation and storage of the mipmap levels for 6 RGBA textures. On our hardware[1], this solution turns out to be slower than a readback and CPU-based summation. Furthermore, using a PCI-Express graphics card and the asynchronous readbacks provided by OpenGL Pixel Buffer Objects [SA08], the pipeline stalls introduced by the readback tend to be quite small. However, due to the fast evolution of graphics hardware, the possibility of using GPU-based summation should not be neglected when starting a new implementation.

5.3.3   IRRADIANCE GRADIENT COMPUTATION

The gradient formulas derived in Section 2.1.1 only apply to the particular case of cosine-proportional hemisphere sampling. Hence, we re-derive new gradient formulas suitable for our GPU hemisphere sampling based on the projection onto a single plane.

[1]A PC with 3.6 GHz Pentium 4 CPU, 1 GB RAM, and nVidia Quadro FX 3400 PCI-E.

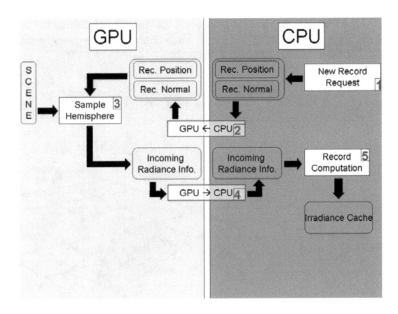

Figure 5.5: New record computation process. The numbers show the order in which the tasks are processed.

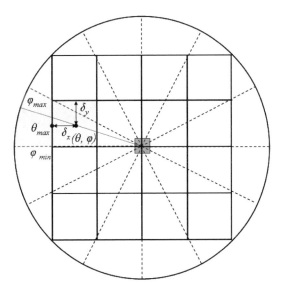

Figure 5.6: Sampling hemisphere seen from above. Edge pixels of the sampling plane are virtually extended to cover the remaining (unsampled) solid angle.

5.3.3.1   Rotation Gradient

Similar to Section 2.1.1.1 (pg. 21), the rotation gradient is computed by summing the marginal gradient contributions from each pixel on the sampling plane. Each contribution is given by differentiating the cosine term $\cos\theta_{j,k}$. The final rotation gradient formula is:

$$\nabla_r E \approx \sum_{k=0}^{N-1}\sum_{j=0}^{N-1} -2L_{j,k}\,\Omega_{j,k}\,\sin\theta_{j,k}\,\hat{v}_{j,k}, \tag{5.13}$$

where $\hat{v}_{j,k}$ is a tangent-plane unit vector perpendicular to the direction $\mathbf{d}_{j,k}$ through the pixel center, i.e.

$$\hat{v}_{j,k} = \frac{(0,0,1)\times\mathbf{d}_{j,k}}{\|(0,0,1)\times\mathbf{d}_{j,k}\|}.$$

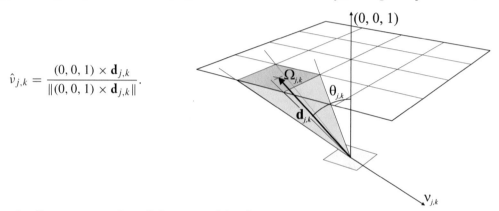

Similar to irradiance computation, all the terms of the above sum except $L_{j,k}$ can be pre-computed and pre-multiplied into a single vector for each pixel.

5.3.3.2   Translation Gradient

The translation gradient derivation proceeds along the same lines as in Appendix B. The resulting gradient formula differs, though, because of the different hemisphere sample distribution.

First, we estimate the differential change in the solid angle of each pixel (j,k) on the sampling plane with translation along the unit vectors $\mathbf{x}$ and $\mathbf{y}$ defining the tangent plane at the record location. For a camera with the field of view of 126.87 degrees, we obtain:

$$\nabla_{\mathbf{x}}\Omega_{j,k} = \frac{4\cos^2\theta_{j,k}}{N}\frac{1}{\min\{r_{j,k},r_{j-1,k}\}}$$

$$\nabla_{\mathbf{y}}\Omega_{j,k} = \frac{4\cos^2\theta_{j,k}}{N}\frac{1}{\min\{r_{j,k},r_{j,k-1}\}},$$

where:

$\Omega_{j,k}$ denotes the solid angle of pixel (j,k),
$\nabla_{\mathbf{x}}\Omega_{j,k}$ is the solid angle derivative with translation along $\mathbf{x}$.
$\theta_{j,k}$ is the polar angle of the direction corresponding to pixel (j,k) (adjusted for the edge pixels),
$N\times N$ is the resolution of the sampling plane in pixels, and
$r_{j,k}$ is the distance to the nearest surface visible through pixel (j,k).

Second, the marginal change of irradiance through each pixel is given by interpolating the radiance from neighboring pixels with the solid angle change used as a blending factor:

$$\begin{aligned}
\nabla_{\mathbf{x}} L_{j,k} &= \nabla_{\mathbf{x}} \Omega_{j,k} (L_{j,k} - L_{j-1,k}), \\
\nabla_{\mathbf{y}} L_{j,k} &= \nabla_{\mathbf{y}} \Omega_{j,k} (L_{j,k} - L_{j,k-1}).
\end{aligned}$$

Finally, we sum up the marginal contributions to get the translation gradient formula:

$$\nabla_t E = \sum_{j=0}^{N-1} \sum_{k=0}^{N-1} \mathbf{x} \cdot \nabla_{\mathbf{x}} L_{j,k} + \mathbf{y} \cdot \nabla_{\mathbf{y}} L_{j,k}. \tag{5.14}$$

Note that the gradient formulas (5.13) and (5.14) express the gradients in the local coordinate frame at the record location. To avoid unnecessary frame transformations during irradiance interpolation, we transform the gradients into the global coordinate frame before storing them in the cache.

5.4   GLOBAL ILLUMINATION RENDERING

The final image is generated in five main steps (Algorithm 17). Given a camera, Step 1 consists in obtaining per-pixel information about visible objects: their position $\mathbf{p}$, surface normal $\mathbf{n}$, and reflectance ρ_d. This information can be efficiently generated in a single pass using Frame Buffers Objects [SA08] and multiple render targets, hence performing vertex processing and rasterization only once.

In Steps 2 to 4, the rendering process determines where new irradiance records are necessary to achieve the user-defined accuracy of indirect illumination computation. In these steps, the irradiance splatting is performed using the GPU or the CPU depending on the current context: in Step 2, each existing record (possibly computed for previous frames) is splatted onto the splat buffer using the GPU. Step 3 consists in reading back the irradiance splat buffer into the CPU memory. In Step 4, the algorithm iterates over the pixels of the irradiance splat buffer to determine where new irradiance records are required (Cache miss detection). If the cumulated weight of a pixel (x, y) in the splatbuffer is zero, a new record is generated at the corresponding position in the scene (i.e. surface point visible through pixel (x, y)). The new record is immediately splatted by the CPU onto the irradiance splat buffer (Figure 5.7).

Once SplatBuff$(x, y).w > 0$ for every pixel, that is to say all pixels are covered by at least one record, the data stored in the cache can be used to display the indirect illumination. At that time in the algorithm, the irradiance cache stored in the CPU memory differs from the cache stored on the GPU: the copy on the GPU represents the cache before the addition of the records described above, while the copy on the CPU is up-to-date. The new records are therefore also added to the cache on the GPU.

Note that no spatial queries are needed in this algorithm; the cache in the CPU memory is a simple linear list. On the GPU, this list is translated into a Vertex Buffer Object [SA08].

The last rendering step is the generation of the final image using the cache contents (Figure 5.8). The irradiance cache on the GPU is updated, then the irradiance splatting algorithm is applied on each newly generated cache record. Hence, the irradiance splat buffer contains the cumulated record weight and outgoing radiance contribution of all the irradiance records. Then, the cumulated contribution at each pixel is divided by the cumulated weight. This process yields an image of the indirect lighting in the scene from the current point of view.

The direct lighting computation is straightforwardly carried out by the GPU, typically using per-pixel lighting and shadow maps [Wil78, BP04]. To reduce the aliasing of shadow maps without harming

---

Algorithm 17 GPU-Based global illumination rendering with irradiance caching

---

// Step 1
Generate position, normal, and reflectance ρ_d of objects visible through pixels (GPU)
Clear the splat buffer
// Step 2
for all existing cache records **do**
　　// The irradiance cache is empty for the first image,
　　// and non empty for subsequent images
　　Algorithm 14: Splat the records onto the irradiance splat buffer (GPU)
end for
// Step 3
Read back the irradiance splat buffer from GPU to CPU
// Step 4
for all pixels (x, y) in the irradiance splat buffer **do**
　　if SplatBuff$(x, y).w = 0$ **then**
　　　　Compute a new record at the corresponding hit point (GPU/CPU)
　　　　Algorithm 14: Splat the *weight* of the new record (CPU)
　　end if
end for
// Step 5
for all cache records **do**
　　Algorithm 14: Splat all the newly generated records (GPU)
end for
// Normalize the irradiance splat buffer
// and compute outgoing direct and indirect radiance (GPU)
for all pixels (x, y) in the irradiance splat buffer **do**
　　SplatBuff$(x, y).E$ /= SplatBuff$(x, y).w$
　　$L_o(x, y) \leftarrow \text{DirectLighting}() + \text{SplatBuff}(x, y).E \cdot \rho_d(x, y)/\pi$
end for

---

the performance, we use high resolution shadow maps (typically 1024×1024) with percentage-closer filtering [RSC87] using 16 samples. Hence, the same shadow map can be used for both the record computation and the rendering of direct lighting in the final image. Note that higher quality images could be obtained using view-dependent shadowing methods such as shadow volumes [Hei91].

5.5   RESULTS

The images and timings in this section have been generated using an nVidia Quadro FX 3400 PCI-E and a 3.6 GHz Pentium 4 CPU with 1 GB RAM.

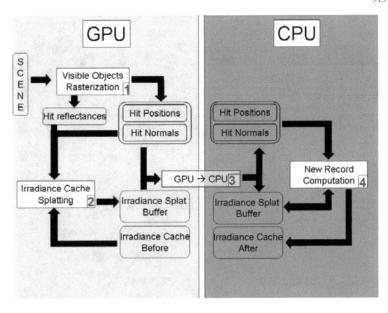

Figure 5.7: The irradiance cache filling process. The numbers show the steps defined in Algorithm 17. During this process, the irradiance cache stored on the CPU is updated whereas the copy on the GPU remains untouched.

5.5.1 HIGH QUALITY RENDERING

This section details non-interactive high quality global illumination computation. First, we compare the results obtained with our GPU-based renderer with the well-known *Radiance* software [War94] in the context of irradiance caching.

We have compared the rendering speed of *Radiance* and our renderer in diffuse environments: the *Sibenik Cathedral* and the *Sponza Atrium* (Figure 5.10). The images are rendered at a resolution of 1000×1000 and use a 64×64 resolution for hemisphere rasterization. The results are discussed hereafter, and summarized in Table 5.2.

a) *Sibenik Cathedral.* This scene contains 80K triangles, and is lit by two light sources. The image is rendered with an accuracy parameter of 0.15. At the end of the rendering process, the irradiance cache contains 4076 irradiance records. The irradiance splatting on the GPU is performed in 188 ms. The Radiance software rendered this scene in 7 min 5 s while our renderer took 14.3 s, yielding a speedup of about $30\times$.

b) *Sponza Atrium.* This scene contains 66K triangles and two light sources. Using an accuracy of 0.1, this image is generated in 13.71 s using 4123 irradiance records. These records are splatted on the GPU in 242.5 ms. Using the Radiance software with the same parameters, a comparable image is obtained in 10 min 45 s. In this scene, our renderer proves about $47\times$ faster than the Radiance software.

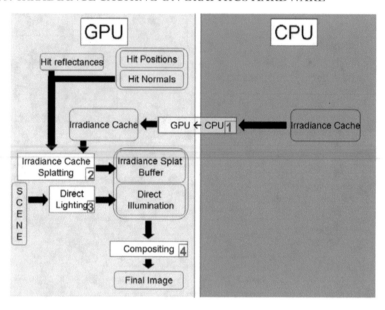

Figure 5.8: The final rendering task. The numbers show the processing order described in Steps 4 and 5 of Algorithm 17.

Table 5.2: Rendering times obtained using radiance and irradiance splatting for high quality rendering of diffuse environments. Each image is rendered at a resolution of 1000×1000.

	Sibenik Cathedral	Sponza Atrium
Triangles	80K	66K
Accuracy a	0.15	0.1
Radiance time (s)	425	645
Our renderer time (s)	14.3	13.7
Speedup	29.7	47.1

5.5.2   INTERACTIVE GLOBAL ILLUMINATION

An important aspect of the irradiance caching algorithm is that the values of the records are view-independent. In a static scene, records computed for a given viewpoint can be reused for other camera positions. Therefore, the irradiance splatting approach can also be used in the context of interactive computation of global illumination using progressive rendering. The direct lighting being computed independently, the user can walk through the environment while the irradiance cache is filled on the fly. Figure 5.11 shows sequential images of *Sam* scene (63K triangles) obtained during an interactive session with an accuracy parameter of 0.5 and resolution of 512×512. The global illumination is computed progressively, by adding at most 100 new records per frame. Our renderer provides an interactive frame rate (between 5 and 32 fps) during this session, allowing the user to move even though the global

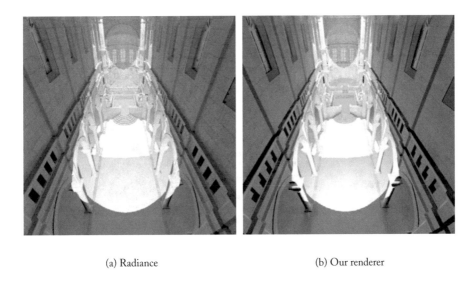

(a) Radiance   (b) Our renderer

Figure 5.9: The Sibenik Cathedral scene (80K triangles). The images show first bounce global illumination computed with Radiance (a) and our renderer (b) (Model courtesy of Marko Dabrovic)

illumination computation is not completed. This method has been used for interactive walkthroughs in diffuse and glossy environments such as the *Sibenik Cathedral* and the *Castle*.

In this context, the irradiance splatting also proves useful for adjusting the value of the user-defined accuracy parameter a. In the classical irradiance caching method, the structure of the octree used to store the records is dependent on a. Therefore, the octree has to be regenerated for each new value of a. Using this approach the size of each splat is computed for each frame on the GPU, hence allowing the user to tune the value of a interactively to obtain the desired visual quality (Figure 5.12).

5.6   CONCLUSION

This chapter presented a reformulation of the irradiance caching algorithm by introducing irradiance splatting. In this method, the validity sphere of each record is splatted onto the image plane. For each pixel within the splatted sphere, a fragment shader computes the contribution of the record to the indirect lighting of the corresponding point. The record weight and weighted contribution are accumulated into the irradiance splat buffer using floating-point alpha blending. The final indirect lighting of visible points is obtained by dividing the weighted sum of the records by the cumulated weights in a fragment shader. Using this approach, each record contributes to all possible pixels. As a side effect this also simplifies the cache miss detection algorithm by allowing a linear traversal of the image. By avoiding the need of complex data structures and by extensively using the power of graphics hardware, the irradiance splatting allows for displaying global illumination in real-time.

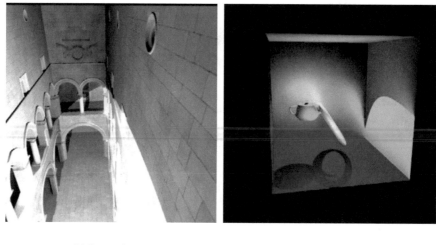

(a) Sponza Atrium   (b) Cornell Box

Figure 5.10: Images obtained with our renderer. The Sponza Atrium (66K triangles) is only made of diffuse surfaces (Model courtesy of Marko Dabrovic). The Cornell Box (1K triangles) contains a glossy back wall, which was rendered with a GPU implementation of radiance caching [KGPB05].

A major speedup factor, however, is the use of both GPU and CPU to compute the values of the irradiance records. While the GPU performs the costly hemisphere sampling, the CPU sums up the incoming radiances to compute the actual values and gradients of the records.

Although we describe irradiance splatting specifically in the context of a GPU implementation of irradiance caching, the scope of the algorithm is more widespread. In particular, irradiance splatting can be straightforwardly extended to radiance splatting for rendering global illumination in glossy environments [GKBP05]. Also, the algorithm has been described in the context of the computation of single-bounce global illumination. However, the splatting operation can be applied recursively when records are generated, yielding multiple layers of irradiance caches as in the Radiance software.

Furthermore, the idea of irradiance splatting can be useful in CPU implementations, too. For example, in their adaptive radiance caching, Křivánek et al. [KBPv06] organize all shading points for an image into a kd-tree. The shading points to which a record could potentially contribute are located using a range query in the kd-tree. With a kd-tree, the shading points are not restricted to correspond to image pixels—points seen through specular reflection or transmission can also be included. The kd-tree based splatting can be useful in relighting engines [TL04], where the set of shading points is fixed and only materials or lighting change during the session.

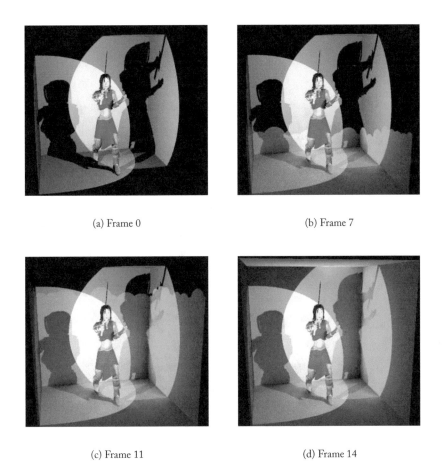

(a) Frame 0   (b) Frame 7

(c) Frame 11   (d) Frame 14

Figure 5.11: A progressive rendering session for interactive visualization of the *Sam* scene (63K triangles). Our renderer computes at most 100 new records per frame, hence maintaining an interactive frame rate (5 fps) during the global illumination computation. When the irradiance cache is full, the global illumination solution is displayed at 32 fps.

(a) $a = 0.5$ (b) $a = 0.2$ using the records generated
                              with $a = 0.5$

(c) $a = 0.2$

Figure 5.12: The irradiance splatting method allows for the dynamic modification of the user-defined accuracy value a, hence easing the parameter tuning process.

CHAPTER 6

Temporal Irradiance Caching

Most global illumination methods only focus on static scenes. Accounting for animated objects and light sources either require a recomputation of the global illumination solution for each frame or involve complex data structures and algorithms for temporal optimization [SKDM05]. Furthermore, global illumination solutions in dynamic scenes commonly exhibit all sorts of temporal artifacts, such as flickering, popping, etc.

This chapter describes a simple yet accurate method for the computation of global illumination in animated environments [GBP07], where the viewer, objects and light sources move. As irradiance caching leverages the spatial coherence of indirect lighting to reduce the cost of global illumination, we consider an extension for sparse sampling and interpolation in time—the temporal irradiance caching.

A simple and commonly used approach to render animations with irradiance caching consists in discarding all the cached records and starting with an empty cache at the beginning of each frame. This indiscriminate discarding of records amounts to significant waste of computational effort. Additionally, the resulting animation may exhibit flickering and popping artifacts since the record positions are likely to fluctuate from frame to frame.

6.1 OVERVIEW OF TEMPORAL IRRADIANCE CACHING

The temporal irradiance caching amortizes the cost of irradiance record computation onto several frames by performing a sparse temporal sampling and temporal interpolation of irradiance (Algorithm 18). Apart from the improved performance, the temporal interpolation also converts the high frequency temporal noise (i.e. the flickering artifacts) into much less noticeable low frequency temporal errors.

When a record i is created at frame n, the future incoming lighting is estimated. This estimate is first used to compute the ratio between the current and future lighting. In the spirit of the spatial caching, we define the temporal weighting function w_i^t as the inverse of the temporal change. Hence the number of frames in which i can be reused is inversely proportional to the irradiance's temporal rate of change.

We use a fast reprojection technique for estimating the future lighting using the data sampled at the current frame.

Finally, similar to the translation and rotation gradients used for irradiance interpolation in space, we introduce temporal gradients for smooth temporal interpolation and extrapolation of the irradiance in time.

To avoid possible flickering problems due to discarding a record, we follow [TMS02] and keep track of the record locations over time. If a record located at point $\mathbf{p}_i$ cannot be reused in current frame, a new record is created at the same location. Note that the location of records remain constant even though they lie on dynamic objects (Figure 6.1).

6.2 TEMPORAL WEIGHTING FUNCTION

The temporal weighting function expresses the confidence in a given record as a function of time. Similar to spatial caching, we define the weighting function as the inverse of the temporal change ϵ^t of irradiance

---

Algorithm 18 Temporal Irradiance Caching

---

    for all frames n **do**
        for all existing records i **do**
            if $w_i^t(n)$ is greater than a threshold **then**
                Use i in frame n
            end if
        end for
        for all points $\mathbf{p}$ where a new record is needed **do**
            Sample the hemisphere above $\mathbf{p}$
            Estimate the future incoming lighting (Section 6.4)
            Generate w_i^t (Section 6.2)
            Compute the temporal gradients (Section 6.3)
            Store the record in the cache
        end for
    end for

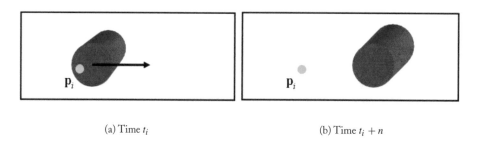

(a) Time t_i (b) Time $t_i + n$

Figure 6.1: Record i created at time t_i remains at point $\mathbf{p}_i$ even though it lays on a dynamic object.

between time t and t_0:

$$\epsilon^t = \frac{\partial E}{\partial t}(t_0)\,(t - t_0). \tag{6.1}$$

The derivative $\frac{\partial E}{\partial t}(t_0)$ can be approximated using estimates of irradiance at two successive times t_0 and t_1, denoted E_0 and E_1.

$$
\begin{aligned}
\frac{\partial E}{\partial t}(t_0) &\approx \frac{E_1 - E_0}{t_1 - t_0} \\
&= \frac{\tau E_0 - E_0}{t_1 - t_0} \quad \text{where } \tau = E_1/E_0 \\
&= E_0 \frac{\tau - 1}{t_1 - t_0}.
\end{aligned}
\tag{6.2}
$$

The time range of the animation is discretized into integer frame indices. Therefore, we always choose $t_1 - t_0 = 1$, i.e. E_1 and E_0 represent the estimated irradiance at two successive frames.

As in the spatial caching we define the temporal weighting function as the inverse of the change, excluding the term E_0:

$$w_i^t(t) = \frac{1}{(\tau - 1)(t - t_0)}. \tag{6.3}$$

where $\tau = E_1/E_0$ is named *temporal irradiance ratio*.

To determine the temporal validity domain of a record, the temporal weighting function is tested against a user-defined *temporal accuracy* value a^t. A record i created at time t_0 is allowed to contribute to the image at time t if

$$w_i^t(t) \geq 1/a^t. \tag{6.4}$$

This way, the temporal validity domain of a record depends on the local variations of irradiance in time. Since a given record can be reused in several frames, the computational cost can be significantly reduced.

Maximum Record Lifespan Equation (6.2) shows that if the environment remains static starting from frame t_0, we obtain $\tau = 1$. Therefore, according to Equation (6.4), w_i^t is infinite for any frame, and hence record i is allowed to contribute at any time $t > t_0$. However, since part of the environment may be dynamic, the inaccuracy becomes significant as $t - t_0$ increases (Figure 6.2). This is a limitation of our technique for estimating the temporal change of irradiance, which determines the lifespan of a record by only considering the local change between E_t and E_{t+1}. To overcome this problem, we use a user-defined value $\delta_{t_{max}}$ to threshold the length of the validity domain associated with each record. If Equation (6.5) does not hold, we decide that the record cannot be reused in frame t.

$$t - t_0 < \delta_{t_{max}}. \tag{6.5}$$

This reduces the risk of using obsolete records which may produce artifacts due to residual global illumination effects also known as "ghosts". In our current implementation, this value must be user-defined by trial and error to obtain the best results. If $\delta_{t_{max}}$ is too low, many records may be recomputed unnecessarily. If it is too high, the same records might be reused in too many frames and ghosting may be an issue.

6.3 TEMPORAL GRADIENTS

The temporal weighting function provides a simple and adaptive way of leveraging temporal coherence by introducing an aging method based on the change of irradiance. Nevertheless replacing obsolete records by new ones creates a discontinuity of irradiance, which causes visible flickering/popping artifacts. Therefore, we introduce *temporal gradients* to generate a smooth and less noticeable transition between successive records.

Temporal gradients are conceptually similar to classical irradiance gradients. Instead of representing the irradiance change with respect to translation and rotation, those gradients represent how the irradiance evolves over time.

In the classical irradiance caching, we use the rotation and translation gradients, $\nabla_{\mathbf{n}} E_i$ and $\nabla_{\mathbf{p}} E_i$,[1] to extrapolate irradiance of record i at a point $\mathbf{p}$ with normal $\mathbf{n}$:

$$E_i(\mathbf{p}) = E_i + (\mathbf{n}_i \times \mathbf{n}) \cdot \nabla_{\mathbf{n}} E_i + (\mathbf{p} - \mathbf{p}_i) \cdot \nabla_{\mathbf{p}} E_i,$$

---

[1]The rotation and translation gradients are denoted $\nabla_r E_i$ and $\nabla_t E_i$ in other chapters of the book. We use $\nabla_{\mathbf{n}} E_i$ and $\nabla_{\mathbf{p}} E_i$ here to avoid the double use of t, for translation and for time.

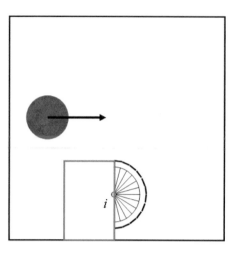
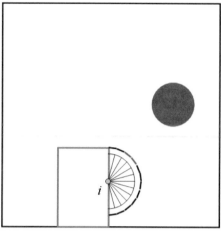

(a) Time t_i (b) Time $t_i + n$

Figure 6.2: When record i is created at time t_i, the surrounding environment is static ($\tau_i = 1$). However, the red sphere is visible from the record i n frames later. The user-defined value $\delta_{t_{max}}$ prevents the record from contributing if $n > \delta_{t_{max}}$, reducing the risk of using obsolete records.

where E_i, $\mathbf{p}_i$, and $\mathbf{n}_i$, are the irradiance, position, and normal of record i, respectively.

We extend this notion of gradient-based irradiance extrapolation to temporal domain by introducing the *temporal irradiance gradient* $\frac{\partial}{\partial t} E$ that gives us a first order approximation of irradiance change in time. Since the rotation and translation gradients themselves also evolve in time, we additionally introduce the temporal gradient of the rotation and translation gradients: $\frac{\partial}{\partial t} \nabla_{\mathbf{n}} E$ and $\frac{\partial}{\partial t} \nabla_{\mathbf{p}} E$.

Using the temporal gradients, the contribution of record i created at time t_i to the irradiance at point $\mathbf{p}$ at time t is estimated by:

$$
\begin{aligned}
E_i(\mathbf{p}, t) \;=\; & E_i + \\
& \frac{\partial}{\partial t} E_i \, (t - t_i) + \\
& (\mathbf{n}_i \times \mathbf{n}) \cdot (\nabla_{\mathbf{n}} E_i + \frac{\partial}{\partial t} \nabla_{\mathbf{n}} E_i (t - t_i)) + \\
& (\mathbf{p} - \mathbf{p}_i) \cdot (\nabla_{\mathbf{p}} E_i + \frac{\partial}{\partial t} \nabla_{\mathbf{p}} E_i (t - t_i)),
\end{aligned}
$$

where

$\frac{\partial}{\partial t} E_i$ is the *temporal gradient of irradiance,*

$\frac{\partial}{\partial t} \nabla_{\mathbf{n}} E_i$ is the *temporal gradient of rotation gradient,* and

$\frac{\partial}{\partial t} \nabla_{\mathbf{p}} E_i$ is the *temporal gradient of translation gradient.*

This formulation represents the temporal change of the irradiance around $\mathbf{p}_i$ as 3 quantities: the actual change of irradiance at point $\mathbf{p}_i$, and the change of the translation and rotation gradients over time. Those quantities can be calculated using the estimate of future irradiance, described in the next section. Since this method estimates the irradiance at time $t+1$ using the information available at time t, we define *extrapolated* temporal gradients as:

$$\begin{aligned}
\frac{\partial E_i}{\partial t} &\approx E_i(t_i+1) - E_i(t_i) \\
\frac{\partial}{\partial t}\nabla_{\mathbf{n}}E_i &\approx \nabla_{\mathbf{n}}E_i(t_i+1) - \nabla_{\mathbf{n}}E_i(t_i) \\
\frac{\partial}{\partial t}\nabla_{\mathbf{p}}E_i &\approx \nabla_{\mathbf{p}}E_i(t_i+1) - \nabla_{\mathbf{p}}E_i(t_i).
\end{aligned} \tag{6.6}$$

However, the extrapolated temporal gradients do not remove all the discontinuities in the animation. When a record i is replaced by record l, the accuracy of the result exhibits a possible discontinuity, yielding some flickering artifacts. As explained in section 6.2, this problem can be avoided by keeping track of the history of the records: when record i gets obsolete, a new record l is created at the same location. Since $t_l > t_i$, we can use the value of irradiance stored in l to compute *interpolated* temporal gradient for record i:

$$\begin{aligned}
\frac{\partial E_i}{\partial t} &\approx (E_l - E_i)/(t_l - t_i) \\
\frac{\partial}{\partial t}\nabla_{\mathbf{n}}E_i &\approx (\nabla_{\mathbf{n}}E_l - \nabla_{\mathbf{n}}E_i)/(t_l - t_i) \\
\frac{\partial}{\partial t}\nabla_{\mathbf{p}}E_i &\approx (\nabla_{\mathbf{p}}E_l - \nabla_{\mathbf{p}}E_i)/(t_l - t_i).
\end{aligned} \tag{6.7}$$

The temporal gradients improve the continuity of irradiance, hence removing the flickering artifacts. However, the gradients only account for the first derivative of the change of incoming lighting, which temporally smoothes the changes. While this method proves accurate in scenes with smooth changes, it should be noted that the gradient-based temporal interpolation may introduce ghosting artifacts when used in scenes with very sharp changes of illumination. In this case, a^t and $\delta_{t_{max}}$ must be reduced to obtain a sufficient update frequency.

As shown in Equation (6.2), the determination of the temporal weighting function and extrapolated gradients relies on the knowledge of the incoming radiance at the next time step, E_{t_0+1}. Since the explicit computation of E_{t_0+1} would introduce a significant computational overhead, we propose a simple and accurate estimation method based on reprojection.

6.4 ESTIMATING FUTURE IRRADIANCE E_{t_0+1}

We use the reprojection and hole-filling approach proposed by Walter *et al.* [WDP99]. While Walter *et al.* use reprojection for interactive visualization using ray tracing, our aim is to provide a simple and reliable irradiance estimate at a given point at time $t_0 + 1$ by using the data acquired at time t_0 only. This estimate will be used to determine the lifespan of the records by evaluating the temporal weighting function.

In the context of a predefined animation, the changes in the scene are known and accessible at any time. When a record i is created at time t_0, the hemisphere above $\mathbf{p}_i$ is sampled (Figure 6.3(a)) to compute the irradiance and gradients at this point. Since the changes between times t_0 and $t_0 + 1$ are

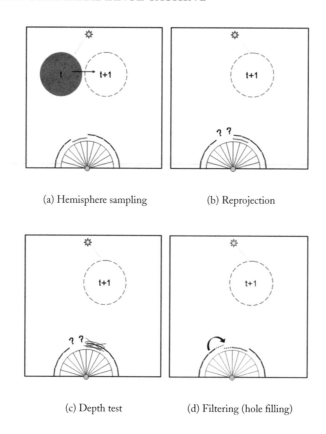

(a) Hemisphere sampling   (b) Reprojection

(c) Depth test   (d) Filtering (hole filling)

Figure 6.3: The hemisphere is sampled at time t as in the classical irradiance caching process (a). For each ray, the algorithm determines where each visible point will be located at time $t + 1$ by reprojection (b). Distant overlapping points are removed using depth test (c), while resulting holes are filled using the farthest neighboring values (d).

known, it is possible to reproject the points visible at time t_0 to obtain an estimate of the visible points at time $t_0 + 1$ (Figure 6.3(b)). The outgoing radiance of reprojected visible points can be estimated by accounting for the rotation and displacement of both objects and light sources. In overlapping areas, a depth test accounts for the occlusion change (Figure 6.3(c)).

However, some parts of the estimated incoming radiance function may be unknown (holes) due to displacement and overlapping of visible objects (Figure 6.3(d)). As proposed in [WDP99], we use a simple hole-filling method: each hole is filled using the background values, yielding a plausible estimate of the future indirect lighting. In practice, the neighboring pixel corresponding to the farthest object is considered as "background", and then directly copied into the hole. As the motion of objects tend to be small between successive frames, a small neighborhood (typically 3×3) is generally sufficient for the background search.

As shown in Figure 6.4 and Table 6.1, the reprojection reduces the error in the estimate of the future incoming lighting. Those errors were measured by comparing the irradiances at time $t+1$ with the irradiances estimated using records created at time t.

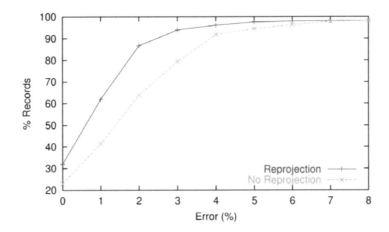

Figure 6.4: Error between the actual lighting at $t+1$ and the lighting at $t+1$ estimated with and without reprojection. This plot represents the percentage of records for which the estimate of the future incoming lighting is below a given RMS error level. The reprojection reduces the overall error in the estimate compared to a method without reprojection (i.e. where the lighting is considered temporally constant). Errors computed using 4523 records.

Table 6.1: RMS error between the actual lighting at $t+1$ and the lighting at $t+1$ estimated with and without reprojection. (Based on 4523 values).

Error	Reprojection	No Reprojection
Min	0%	0%
Max	30%	32%
Mean	2.9%	3.7%
Median	1.6%	2.4%

6.5   GPU IMPLEMENTATION

In this section, we discuss some GPU implementation details of temporal caching for increased performance. First, we detail the implementation of the future incoming radiance estimate by reprojection. Then, we describe how the GPU can be simply used in the context of irradiance cache splatting (Chapter 5) to discard useless records and avoid their replacement.

Reprojection of Incoming Radiance   As shown in section 6.2, the computation of the temporal weighting function and temporal gradients for a given record i require an estimate of the irradiance reaching point $\mathbf{p}_i$ at the next time step. This estimate is obtained through reprojection (section 6.4), provided that the position of the objects at next time step is known. Therefore, for a given vertex v of the scene and a given time t, the transformation matrix corresponding to the position and normal of v at time $t + 1$ is known. Such matrices are available for light sources as well. Using the method described in Chapter 5, the irradiance value of a record i can be generated by rasterizing the scene on a single plane above point $\mathbf{p}_i$. The reprojection is carried out in three passes on the GPU using specific shaders.

In a first pass, during the rasterization at time t, the fragment shader outputs information on the current and future lighting situations: on the one hand, the fragment shader computes the current lighting at the points visible through each fragment. On the other hand, it also estimates the location of each of those visible points at time $t + 1$, as well as the corresponding lighting. The generated set of future positions and their associated radiances can then be considered as a point cloud approximating the geometry surrounding the record location.

As mentioned above, the regular way of estimating the incoming radiance on the GPU is by rasterizing the scene and computing the incoming radiance at each visible point. In the case of the estimation of the future incoming radiance function, we avoid sampling the actual scene by rasterizing the point cloud generated in the first pass instead. Therefore, in the second pass, each of those vertices is sent to the graphics pipeline as a pixel-sized point. The result of the rasterization process is an estimate of the incoming radiance function at time $t + 1$. Since the size of the sampling plane is usually small (typically 64×64), this process is generally much faster than resampling the whole scene.

However, the point cloud is not as well-behaved as the original, polygonal scene. In particular, the point cloud contains exactly the same number of points as the number of pixels in the sampling image. Therefore, while the overlapping of points can be simply solved by classical z-buffer, the resulting image may contain holes at the location of dynamic objects (Figure 6.3(d)).

Since the time shift between two successive frames is very small, the holes also tend to be small. As described in Section 6.4, we use a third pass to fill the holes by extrapolation using the local background (that is, the neighboring value with the highest depth). This computation can be performed efficiently on the GPU using the stencil buffer with an initial value of 0. During the reprojection process, each rasterized point increments the stencil buffer. Therefore, the hole-filling algorithm must be applied only on pixels where the stencil buffer is still 0, that is where no point have been mapped. The final result of this algorithm is an estimate of the irradiance at time $t + 1$, generated entirely on the GPU (Figure 6.5). This estimate is used in the extrapolated temporal gradients and the temporal weighting function. As shown in Equation (6.4), this latter defines a maximum value of the lifespan of a given record, and triggers its recomputation. However, this recomputation is not always necessary.

Replacement/Deletion Method   Therefore, if a record cannot contribute to the current image (i.e. out of the view frustum, or occluded), it can be simply deleted instead of being replaced by a novel, up-to-date record. This avoids the generation and update of a "trail" of records following dynamic objects (Figure 6.6), hence reducing the memory and computational costs. In the context of the irradiance cache splatting [GKBP05] described in Chapter 5, this decision can be easily made using hardware occlusion queries: during the last frame of the lifespan of record i, an occlusion query is issued as the record is rasterized. In the next frame, valid records are first rendered. If a record i is now obsolete, the result of the occlusion query is read from the GPU. If the number of covered pixels is 0, the record is discarded. Otherwise, a new record l is computed at location $\mathbf{p}_l = \mathbf{p}_i$.

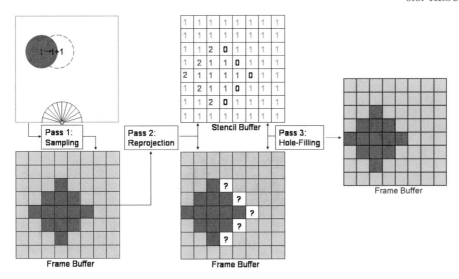

Figure 6.5: Reprojection using the GPU. The first pass samples the scene to gather the required information. Then, the visible points are reprojected to their estimated position at next time step. During this pass, each rendered fragment increments the stencil buffer. Finally, the holes (i.e. where the stencil value is 0) are filled using the deepest neighboring values.

The hardware occlusion queries are very useful, but they suffer from high latency. However, in this method, the result of a query is not needed immediately. Between the query issue and the reading of the record coverage, the renderer renders the other records, then switches the scene to next frame and renders valid records. In practice, the average latency is negligible (less than 0.1% of the overall computing time). In our test scenes this method reduces the storage and computational costs by up to 25-30%.

6.6   RESULTS

This section discusses the results obtained using temporal irradiance caching and compares them with the classical method in which a new cache is computed for each frame. This latter method is referred to as *per-frame computation* in the remainder of this section. The images, videos and timings have been generated using a 3.8GHz Pentium 4 with 2 GB RAM and an nVidia GeForce 7800 GTX 512MB. The scene details and timings are summarized in Table 6.2.

Cube in a Box   This very simple, diffuse scene (Figure 6.10(a)) exhibits high flickering when no temporal gradients are used. Along with a significant speedup, the temporal caching reduces the flickering artifacts by using extrapolated temporal gradients. The artifacts become unnoticeable with interpolated gradients. The animations are generated using $a^t = 0.05$, and a maximum lifespan of 20 frames. The per-frame computation requires 772K records to render the animation. In temporal caching, only 50K records are needed, yielding a memory load of mere 12.4 MB. Figure 6.7 shows the accuracy values obtained with and without temporal gradients. The remaining flickering of the extrapolated temporal gradients are due

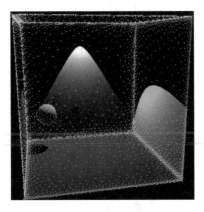

(a) Time 1

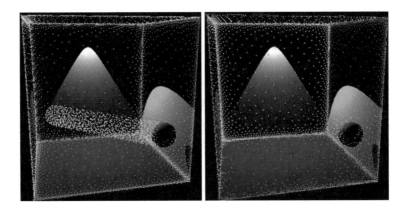

(b) Time 100, systematic update   (c) Time 100, record removal

Figure 6.6: The sphere moves from the left to the right of the box. At time 1 (a), records (represented by green points) are generated to compute the global illumination solution. When the sphere moves, new records are created to evaluate the irradiance on the sphere. If every record is permanently kept up-to-date, a "trail" of records lies on the path of the dynamic sphere (b). Using record removal, only useful records are updated (c).

to the discontinuities of accuracy. Since our aim is high quality rendering, the following results focus on interpolated temporal gradients which avoid discontinuities.

Moving Light   A similar scene (Figure 6.10(b)) illustrates the behavior of the algorithm in animations with dynamic light sources. The bottom of the box is tiled to highlight the changes of indirect lighting.

Table 6.2: Test scenes and timings for temporal caching.					
Scene	# of polys	# of frames	Per-frame comp.	Temporal caching	Speedup
Cube in a Box	24	400	2048 s	269 s	7.62
Moving Light	24	400	2518 s	2650 s	0.95
Flying Kite	28K	300	5109 s	783 s	6.52
Japanese Interior	200K	750	13737 s	7152 s	1.90
Spheres	64K	200	3189 s	753 s	4.24

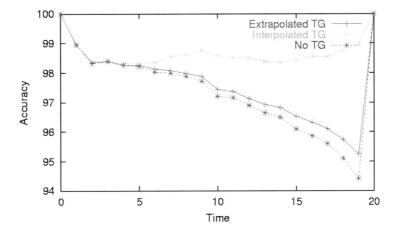

Figure 6.7: Plot of the temporal accuracy as a function of time obtained in scene Cube in a Box by creating records at time 0 and extrapolating their value until time 19. The accuracy value at a time t is obtained by computing the error between the extrapolated values and the actual lighting at time t. At time 20, the existing records are discarded and replaced by up-to-date records with maximum accuracy. The temporal gradients (TG) provide a better approximation compared to the approach without those gradients. Using interpolated gradients, the accuracy is continuous and remains above 98%.

Due to the highly dynamic indirect lighting, the lifespan of the records is generally very short, yielding frequent updates of irradiance values. Compared to per-frame computation, the temporal caching scheme allows to render the animation with higher quality in a comparable time. The small overhead of temporal radiance caching is due to the estimation of the record lifespan: in this scene, most of the records have a lifespan of 1 frame due to the rapidly changing lighting conditions. Therefore, compared to per-frame computation, the temporal caching tends to perform more computations, yielding a higher temporal quality at the detriment of computational efficiency.

Flying Kite   In a more complicated, textured scene (Figure 6.10(c)), the algorithm also provides a significant quality improvement while drastically reducing the computation time. In the beginning of the animation the change of indirect lighting is small, and hence the records can be reused in several frames. However, when the kite gets down, its dynamic reflection on the ceiling and wall is clearly noticeable. Using the temporal weighting function, the global illumination solution of this zone is updated at a fast pace, avoiding ghosts in the final image (Figure 6.8).

(a) Actual sampling frame   (b) Records lifespan

Figure 6.8: When computing the global illumination solution for the current frame (a), the temporal caching estimates where the lighting changes. The lifespan of each generated record is computed by estimating the future change of lighting (b). Green and red colors respectively represent long and short lifespan.

Japanese Interior   In this more complex scene (Figure 6.10(d)), the glossy objects are rendered using the radiance caching algorithm [KGPB05]. The animation illustrates some key features of the algorithm: dynamic nondiffuse environment and important changes of indirect lighting. In the beginning of the animation, the scene is lit by a single, dynamic light source. In this case, temporal gradients suppress the flickering artifacts present in per-frame computation but do not provide a significant speedup (1.25×). In the remainder of the animation, most of the environment is static, even though some dynamic objects generate strong changes in the indirect illumination. The temporal gradients take advantage of this situation by adaptively reusing records in several frames. The result is the elimination of flickering and a significant speedup (1.9× on average in our walkthrough animation) compared to per-frame computation. During the generation of this animation, the average latency introduced by occlusion queries is 0.001% of the overall rendering time.

Spheres   This scene features complex animation with 66 diffuse and glossy bouncing spheres and a glossy back wall (Figure 6.10(e)). The temporal caching eliminates the flickering while reducing the computational cost by a factor of 4.24. We used a temporal accuracy value $a^t = 0.05$ and a maximum record lifespan $\delta_{t_{max}} = 5$.

Comparison with Monte Carlo Path Tracing   As the irradiance caching algorithms introduce low frequency spatial errors, the temporal caching algorithm introduces low frequency temporal errors. Therefore, the obtained images contain both spatial and temporal errors. In a sequence of 100 images of the *Cube in a Box* scene, the average RMS error of the irradiance between temporal irradiance caching and the reference solution is 0.139% (Figure 6.9). Even though the results obtained exhibit differences compared to the reference solution, the images obtained are a reliable estimate of the global illumination solution.

(a) Reference   (b) Temporal irrad. caching   (c) Difference

Figure 6.9: Images obtained using Monte Carlo path tracing with 16K rays per hemisphere at each pixel (a) and the temporal irradiance caching algorithm (b). (c) is the difference image of (a) and (b) multiplied by 5 to highlight the differences.

Computational Overhead of Reprojection   During the computation of a record, the temporal caching evaluates the value of the incoming lighting for both current and next time steps. As shown in Section 6.4, the estimation of the future incoming lighting is performed by simple reprojection. Therefore, the related computational overhead is independent of the scene geometry. In our tests, each record was computed at resolution 64×64. On our system, the reprojection is performed in approximately 0.46 ms. For comparison, the time required to compute the actual incoming lighting at a given point in our 200K polygons scene is 4.58 ms. In this case, the overhead due to the reprojection is only 10% of the cost of the actual hemisphere sampling. Even though this overhead is not negligible, this estimate enables us reduce the overall rendering time by reusing the records in several frames.

6.7   CONCLUSION

This chapter presented a method for exploiting temporal coherence in the context of irradiance caching. We introduced an approach for sparse sampling and interpolation of the incoming radiance in the temporal

(a) Cube in a Box   (b) Moving Light   (c) Flying Kite

(d) Japanese Interior   (e) Spheres

Figure 6.10: Images of scenes discussed in Section 6.6.

domain. We defined a temporal weighting function and temporal gradients, allowing a simple and accurate temporal interpolation of incoming radiance values. The results show both a significant speedup and an increased temporal quality compared to per-frame computation. Due to the sparse temporal sampling, the irradiance values for the entire animation segment can be stored within the main memory.

This method lays the basis for simple global illumination in dynamic scenes but leaves many aspects for future work. This would include the design of a more accurate estimation method for extrapolated temporal gradients. Such a method will find use in on-the-fly computation of indirect lighting during interactive sessions. Another improvement would consist in designing an efficient method for faster aging of the records located near newly created records for which important changes have been detected. This would avoid the need for a user-defined maximum validity time, while guaranteeing the absence of global illumination ghosting.

Mathematical Foundations

This appendix summarizes some basic mathematical tools useful (although not strictly required) for understanding the material discussed in the book: calculations on the (hemi)sphere, probability, and Monte Carlo integration.

As this chapter only provides a minimal introduction to those concepts, the reader may want to refer to more thorough descriptions of the mathematics underlying physically-based rendering, such as [Gla95]. Also, a collection of mathematical tools and formulas for global illumination is provided in [Dut03].

A.1 (HEMI)SPHERES: CARE AND FEEDING

Lighting simulation involves many calculations on the sphere or hemisphere, based on the notions of spherical coordinates and solid angles, described below.

A.1.1 SPHERICAL COORDINATES

The location of any 3D point is most often represented using its cartesian coordinates (x, y, z). However, in the context of lighting simulation it is also useful to represent this point using its spherical coordinates (r, θ, ϕ), depicted in Figure A.1. Here

- r is the radius, i.e. the distance from the origin,

- θ is the polar angle, measured from the z axis down, and

- ϕ is the azimuthal angle, measured from the x axis counterclockwise.

On a sphere $(\theta, \phi) \in [0, 2\pi] \times [0, \pi]$ while on the upper hemisphere $(\theta, \phi) \in [0, 2\pi] \times [0, \frac{\pi}{2}]$. Radius $r \in [0, +\infty]$.

A.1.2 DIRECTION

A direction in 3D can be expressed by a unit-length vector in cartesian coordinates or using a pair (θ, ϕ) in spherical coordinates. The conversion from spherical to cartesian coordinates is given by:

$$\begin{aligned} x &= \sin\theta \cos\phi \\ y &= \sin\theta \sin\phi \\ z &= \cos\theta. \end{aligned} \tag{A.1}$$

A.1.3 SOLID ANGLE

Solid angles extend the notion of classical angles from two to three dimensions. In 2D, the angle α subtended by an object with respect to a position $\mathbf{p}$ can be obtained by projecting the object onto a unit circle centered at $\mathbf{p}$. The value of α, measured in radians, is then the length of the arc covered by the projection of the object (Figure A.2(a)). The angle subtended by a circle around $\mathbf{p}$ is 2π radians.

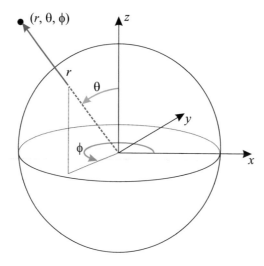

Figure A.1: Spherical coordinates.

Following the same principle, the solid angle Ω subtended by a 3D object with respect to a position **p** is obtained by projecting the 3D object onto a unit sphere centered at **p** (Figure A.2(b)). The unit for solid angles is the steradian, abbreviated sr. The solid angle subtended by the entire sphere is 4π steradians.

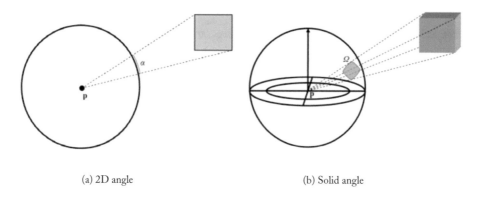

(a) 2D angle   (b) Solid angle

Figure A.2: Angle subtended by an object. In 2D (a), the angle is given by the projection of the object onto the unit circle. In 3D (b), the solid angle is given by the projection of the object onto the unit sphere.

A.1.4   DIFFERENTIAL SOLID ANGLE

Just like the integration of a 1D function $f(x)$ considers infinitely small intervals dx around a value x, the integration of a function over the (hemi)sphere considers infinitely small sets of directions around a main direction ω. This set represents an infinitely small solid angle $d\omega$, called *differential solid angle* (Figure A.3).

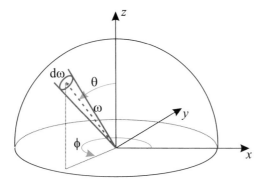

Figure A.3: Differential solid angle.

A.1.5   SPHERICAL INTEGRATION

The integral of a spherical function f over all the directions on the unit sphere is written as:

$$I_f = \int_\Omega f(\omega)d\omega.$$

Using the Fubini's theorem, we can change this multi-dimensional integral into two nested one-dimensional integrals in the spherical coordinates. The relation between a differential solid angle and differential spherical coordinates is:

$$d\omega = \sin\theta \, d\theta \, d\phi.$$

Then the integral becomes:

$$I_f = \int_{\phi=0}^{2\pi} \int_{\theta=0}^{\pi} f(\theta, \phi) \sin\theta \, d\theta \, d\phi.$$

For an integral on the hemisphere, the upper limit for θ is $\pi/2$. As an example, let us calculate the integral of $\cos\theta$ over the upper hemisphere:

$$
\begin{aligned}
\int_{H^+} \cos\theta \, d\omega &= \int_{\phi=0}^{2\pi} \int_{\theta=0}^{\pi/2} \cos\theta \, \sin\theta \, d\theta \, d\phi \qquad \left| \, \text{subst.} : \begin{array}{ccc} t &=& \cos\theta \\ dt &=& -\sin\theta \, d\theta \end{array} \right| \\
&= \int_{\phi=0}^{2\pi} \int_{t=1}^{0} -t \, dt \, d\phi \\
&= \pi. \tag{A.2}
\end{aligned}
$$

A.2 CONTINUOUS RANDOM VARIABLES IN A NUTSHELL

Since the Monte Carlo integration, described in the next section, is based on probabilities, we review some terms and ideas from the probability theory assuming the reader has a basic knowledge on this topic. For a more complete introduction to probability, the reader may refer to a textbook such as [Ros06].

A.2.1 RANDOM VARIABLES AND PROBABILITY DENSITY FUNCTIONS

A *continuous random variable* X is a random value drawn from a *domain* D_X. As an example, consider shooting on a target. The distance of the hit from the target center is our continuous random variable. The domain D_X would be an interval of real numbers starting from zero and going to some maximum distance, $D_X = [0, d_{max}]$. The distance for one shot is a *realization*, or an *observed value* of the random variable.

The *probability density function*, or PDF, $p_X : D_X \to \mathbb{R}^+$ tells us how probable are different observed values of a random variable. ($\mathbb{R}^+$ denotes non-negative real numbers.) More precisely, for a 1D random variable, the integral

$$\int_\alpha^\beta p_X(x)\,dx = \Pr(X \in [\alpha, \beta]),$$

gives us the probability that the observed value is between α and β. For a multi-dimensional random variable, we integrate over a subdomain of D_X.

It directly follows that the probability of a fixed single value is zero. This may be strange at the first sight. If we shoot at a target, the hit is certainly at some particular distance, so how can the probability of this happening be zero? Well, this is actually not what we mean. What we say is that if we picked a certain distance d à priori and then asked someone to shoot, the probability that the actual distance will be equal to what we picked before is zero. This property of continuous random variables has important consequences in lighting simulation for some singular cases, such as point light sources and ideal specular (mirror) reflection.

A PDF always integrates to 1 over the entire domain

$$\int_{D_X} p_X(\mathbf{x})\,d\mathbf{x} = 1,$$

since an observed value will certainly lie in D_X.

A.2.2 EXPECTED VALUE AND VARIANCE

The *expected value* of a continuous random variable X, denoted $E[X]$, is given by

$$E[X] = \int_{D_X} \mathbf{x}\, p(\mathbf{x})\,d\mathbf{x}. \tag{A.3}$$

Suppose we transform the random variable by a function $f : D_X \to \mathbb{R}$, i.e. we pass every observed value of X through f. Then, the expected value of the transformed random variable is:

$$E[f(X)] = \int_{D_X} f(\mathbf{x})\, p(\mathbf{x})\,d\mathbf{x}. \tag{A.4}$$

The *variance* of a random variable is the expected deviation from the expected value:

$$\begin{aligned}
V[X] &= E[(X - E[X])^2] \\
&= E[X^2] - E^2[X].
\end{aligned}$$

(To derive the latter expression, we use $E[(X - E[X])^2] = E[X^2] - 2E[XE[X]] + E^2[X]$. Using the fact that $E[aX] = aE[X]$ and rearranging gives the desired form.)

The *standard deviation* of a random variable is the square root of its variance:

$$\sigma[X] = \sqrt{V[X]}.$$

A.3 MONTE CARLO INTEGRATION IN A NUTSHELL

Lighting simulation involves the integration of functions which are not defined analytically, such as the incoming radiance at a point as a function of the direction ω_i. Therefore, the integration has to be performed numerically. Monte Carlo integration estimates the value of an integral by evaluating the integrand at a set of randomly selected points in the integration domain. Unlike finite elements, this numerical method has the advantage of allowing the integration of functions irrespective of their dimensionality or continuity properties.

A.3.1 MONTE CARLO ESTIMATORS

A Monte Carlo estimator is the core of Monte Carlo integration. Consider the integration of a function $f(\mathbf{x})$ over a domain D:

$$I = \int_D f(\mathbf{x}) d\mathbf{x}.$$

Given a set of N independent random variables $X_1, \ldots, X_N$ identically distributed on D, with the associated probability density function $p(\mathbf{x})$, a Monte Carlo estimator for I is:

$$\hat{I} = \frac{1}{N} \sum_{i=1}^{N} \frac{f(X_i)}{p(X_i)}.$$

Estimator Unbiasedness   If $p(\mathbf{x}) \neq 0$ whenever $f(\mathbf{x}) \neq 0$, the expected value of the above estimator is equal to the value of the integral itself:

$$
\begin{aligned}
E[\hat{I}] &= \frac{1}{N} \sum_{i=1}^{N} E\left[\frac{f(X_i)}{p(X_i)}\right] \\
&= \frac{1}{N} \sum_{i=1}^{N} \int_D \frac{f(\mathbf{x})}{p(\mathbf{x})} p(\mathbf{x}) d\mathbf{x} \\
&= \int_D f(\mathbf{x}) d\mathbf{x}.
\end{aligned}
$$

In short

$$E[\hat{I}] = I. \tag{A.5}$$

An estimator for which Equation (A.5) holds is called *unbiased*. In general, the quantity $|E[\hat{I}] - I|$ is called the *bias*. The bias of an unbiased estimator is zero.

Notice that we estimate the integral by evaluating the integrand at a number of randomly placed points. Each of these evaluations is called a *sample* of the integrand.

Estimator Variance   As the value of $\hat{I}$ is computed using a limited number of samples, we expect some error in the result. For an unbiased estimator, there is no systematic error and the only error is due to the estimator variance. More precisely, the error is proportional to the standard deviation, $\sigma[\hat{I}]$. It is natural to expect more accurate results for an increased number of samples. To analyze the error in terms of the number of samples, let us derive the expression for the estimator variance [Vea97].

Let $Y_i = f(X_i)/p(X_i)$. The estimator for I is then $\hat{I} = \frac{1}{N} \sum_{i=1}^{N} Y_i$. The variance of a single-sample estimate is

$$V[\hat{I}|_{N=1}] = V[Y_1] = E[Y_1^2] - E^2[Y_1] = \int_D \frac{f^2(\mathbf{x})}{p(\mathbf{x})} d\mathbf{x} - I^2,$$

since the expected value of the estimator is the integral itself, i.e. $E[\hat{I}] = E[Y_1] = I$. For $N > 1$ samples, we have

$$V[\hat{I}] = V\left[\frac{1}{N} \sum_{i=1}^{N} Y_i \right] = \frac{1}{N^2} \sum_{i=1}^{N} V[Y_i] = \frac{1}{N} V[Y_1].$$

(We used the following identities: $V[aX] = a^2 V[X]$, and $V[\sum X_i] = \sum V[X_i]$ for X_i independent.) In other words, variance decreases linearly with the number of samples, and *the standard deviation, i.e. the error, decreases with the square root of N*. To decrease the error twice, we have to use four times more samples. Other techniques, such as stratification or importance sampling, can reduce estimator variance without the cost of additional samples.

A.3.2   IMPORTANCE SAMPLING: A VARIANCE REDUCTION TECHNIQUE

In lighting simulation the integrand is most often the incoming radiance function, the evaluation of which involves costly ray tracing. Therefore, it is beneficial to seek other techniques for variance reduction than a simple increase of the number of samples. One of the most commonly used variance reduction techniques in lighting simulation is *importance sampling*.

Consider a PDF $p(\mathbf{x})$ exactly proportional to the integrand $f(\mathbf{x})$, i.e.

$$p(\mathbf{x}) = cf(\mathbf{x}),$$

where the normalization value c is given by

$$c = \frac{1}{\int_D f(\mathbf{x}) \, d\mathbf{x}},$$

so that the PDF integrates to 1 over the domain D. With such a PDF, the variance of the estimator is zero, that is to say, the estimator gives an exact result. Unfortunately, we need to know $\int_D f(\mathbf{x}) \, d\mathbf{x}$ for this to work, which is exactly the value of the integral that we are seeking to find.

Nevertheless, the idea of importance sampling is to use a PDF which is roughly proportional to an *à priori* knowledge of the integrand. Using such a PDF, regions with high values would be sampled more densely, hence capturing most of the integral contributions (Figure A.4(a)).

However, note that a poor choice of the PDF may yield disastrous results as the important parts of the integrand may be poorly sampled, or not sampled at all (Figure A.4(b)).

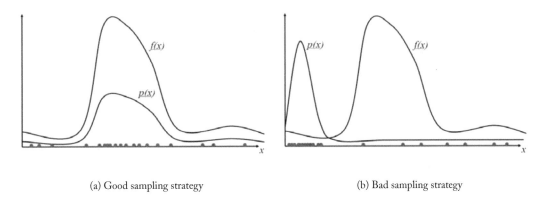

(a) Good sampling strategy   (b) Bad sampling strategy

Figure A.4: When integrating $f(x)$, the PDF $p(x)$ determines the location of the samples (red dots) in the integration domain. When the PDF is well-chosen (a), more samples are generated in the areas where the integrand $f(x)$ has high values, hence reducing the variance. However, using an inappropriate PDF may result in high variance as the sampling may miss some important peaks of the integrand (b).

APPENDIX B

Derivation of Gradient Formulas

This Appendix gives the derivation of formulas for rotation and translation gradients introduced in Section 2.1.1.

B.1 ROTATION GRADIENT DERIVATION

The rotation gradient is estimated from the incoming radiance samples $L_{j,k}$ and their directions $\omega_{j,k}$ used to estimate a new irradiance value at a point. The idea is to compute the contribution to the gradient for each sample separately and then sum the contributions weighted by the incoming radiance to get the resulting gradient estimate. To compute the contribution of one sample, we examine how the cosine term changes with the rotation around a base-plane vector $\mathbf{v}_k$ perpendicular to the sample cell center projected to the base plane, as shown in Figure 2.5, pg. 23. The cosine factor change is given by its derivative with respect to θ, i.e. $-\sin\theta_j$.

Since we are using the same set of radiance samples, the general form of the Monte Carlo estimator for the rotation gradient is nearly identical to the estimator for irradiance:

$$\nabla_r E \approx \frac{1}{MN} \sum_{k=0}^{N-1} \sum_{j=0}^{M-1} \frac{g_{j,k}}{p_{j,k}}.$$

The samples of the integrand g are given by

$$g_{j,k} = -\sin\theta_j L_{j,k} \mathbf{v}_k.$$

Since the samples are still distributed proportionally to the cosine term, the PDF is

$$p_{j,k} = \frac{\cos\theta_j}{\pi}.$$

This yields the final *rotation gradient estimator* (2.4).

B.2 TRANSLATION GRADIENT DERIVATION

Our approach to estimating translation gradient is similar to the procedure we use to estimate the rotation gradient: first we calculate the contribution to the gradient for each hemisphere cell individually and then we sum up the contributions to get the total gradient estimate. The notation used for gradient estimation is illustrated in Figure 2.5 and summarized on page 23.

The stratified sampling strategy used in the irradiance estimator (2.2) divides the hemisphere into cells of solid angle (or area on the unit hemisphere), given by

$$
\begin{aligned}
A_{j,k} &= \int_{\Omega_{j,k}} d\omega \\
&= \int_{\phi_{k_-}}^{\phi_{k+}} \int_{\theta_{j_-}}^{\theta_{j+}} \sin\theta \, d\theta \, d\phi, \\
&= (\cos\theta_{j_-} - \cos\theta_{j_+})(\phi_{k_+} - \phi_{k_-}).
\end{aligned}
\tag{B.1}
$$

To estimate the translation gradient for cell (j, k), we observe how the cell area changes with respect to the displacement of the hemisphere center along two near perpendicular vectors, $\mathbf{u}_k$ and $\mathbf{v}_{k_-}$ (see Figure 2.5). The displacement along $\mathbf{u}_k$ causes a shift of the wall separating the considered cell (j, k) and its neighboring cell $(j - 1, k)$. The induced change of the cell area is given by the directional derivative $\nabla_{\mathbf{u}_k} A_{j,k}$ (i.e. a scalar derivative of $A_{j,k}$ with translation along $\mathbf{u}_k$). We apply the chain rule to find this derivative:

$$
\begin{aligned}
\nabla_{\mathbf{u}_k} A_{j,k} &= \nabla_{\mathbf{u}_k} \theta_{j_-} \cdot \frac{\partial A_{j,k}}{\partial \theta_{j_-}} \\
&= \frac{-\cos\theta_{j_-}}{\min\{r_{j,k}, r_{j-1,k}\}} \cdot \sin\theta_{j_-} (\phi_{k_+} - \phi_{k_-}) \\
&= \frac{2\pi}{N} \frac{\cos\theta_{j_-} \sin\theta_{j_-}}{\min\{r_{j,k}, r_{j-1,k}\}},
\end{aligned}
$$

The derivative $\partial A_{j,k}/\partial\theta_{j_-} = \sin\theta_{j_-}(\phi_{k_+} - \phi_{k_-})$ follows directly from Equation (B.1). The derivative $\nabla_{\mathbf{u}_k} \theta_{j_-} = -\cos\theta_{j_-}/\min\{r_{j,k}, r_{j-1,k}\}$ can be derived by analyzing the geometry of the hemisphere cross-section viewed from a direction perpendicular to $\mathbf{u}_k$, as shown above. Here, $r_{j,k}$ denotes the distance from the hemisphere center to the nearest surface in the sample direction $(\theta_{j,k}, \phi_{j,k})$. It is the distance to the closer of the surfaces seen through two neighboring hemisphere cells, (j, k) and $(j - 1, k)$ that determines the relative movement of the boundary between the two cells. This is why we use the $\min\{r_{j,k}, r_{j-1,k}\}$ term (instead of simply taking $r_{j,k}$) in the estimation of the cell area change. Doing so accounts for the changes in occlusion induced by the translation as pointed out by Ward and Heckbert [WH92].

Similarly, the displacement along $\mathbf{v}_{k_-}$ causes a shift of the wall separating the considered cell (j, k) and its neighboring cell $(j, k - 1)$. Again, we use the chain rule to find the induced cell area derivative:

$$\nabla_{\mathbf{v}_{k\_}} A_{j,k} = \nabla_{\mathbf{v}_{k\_}} \phi_{k_-} \cdot \frac{\partial A_{j,k}}{\partial \phi_{k_-}}$$

$$= \frac{-1}{\sin \theta_{j,k} \min\{r_{j,k}, r_{j,k-1}\}} \cdot (\cos \theta_{j_+} - \cos \theta_{j_-}).$$

The derivative $\partial A_{j,k}/\partial \phi_{k_-} = (\cos \theta_{j_+} - \cos \theta_{j_-})$ follows from Equation (B.1). The directional derivative $\nabla_{\mathbf{v}_k} \phi_{k_-} = -1/(\sin \theta_{j,k} \min\{r_{j,k}, r_{j,k-1}\})$ can be derived by looking at the hemisphere from the top, as shown above.

The change of incoming radiance arriving at the hemisphere center through the considered cell is given by interpolating the radiance from two neighboring cells with the area change used as a blending factor:

$$\nabla_{\mathbf{u}_k} L_{j,k} = \nabla_{\mathbf{u}_k} A_{j,k} (L_{j,k} - L_{j-1,k}),$$
$$\nabla_{\mathbf{v}_{k\_}} L_{j,k} = \nabla_{\mathbf{v}_{k\_}} A_{j,k} (L_{j,k} - L_{j,k-1}).$$

The final translation gradient formula (2.5) is given by summing the marginal irradiance gradients over all hemisphere cells, weighted by the cosine term evaluated at the cell's boundary.

Discussion.   The derivation of the translation gradient presented here is based on the paper [KGBP05] by Křivánek et al., which generalizes the original gradient derivation of Ward and Heckbert [WH92]. As shown by Jarosz [Jar08], the two gradient formulas are equivalent. However, the derivation of Ward and Heckbert, based on an analysis of the projection of hemisphere cells into the base plane, is only applicable to cosine-proportional hemisphere sampling, as described thus far. The derivation presented here, on the other hand, makes it possible to derive translational gradient for an arbitrary distribution of stratified samples. We take advantage of this property to find a gradient formula in the GPU implementation of irradiance caching (Section 5.3.3), where the sample distribution is significantly different.

Another alternative formula for translation gradient has been derived independently by Annen et al. [AKDS04] and Křivánek et al. [KGPB05]. Their formula has the advantage that it does not rely on any stratification so it can be used for an arbitrary hemisphere sampling pattern, such as low-discrepancy sequences used in quasi-Monte Carlo sampling [PH04]. However, their gradient formula does not take occlusion into account which is why it usually produces less accurate gradient estimates.

In the gradient derivation given above, we have assumed that the incoming radiance $L_{j,k}$ does not change with translation of the hemisphere center. This assumption holds only if the surfaces that contribute indirect illumination are purely diffuse. However, the assumption breaks in scenes with participating media and glossy surfaces. This problem is addressed by Jarosz et al. in [JZJ08].

APPENDIX C

Split-Sphere Model

This Appendix details the derivation of Equations (2.6) and (2.8) used for irradiance interpolation. Our goal is to predict the interpolation error caused by reusing a cached value at a different location. The idea proposed by Ward et al. [WRC88] is to derive an upper bound on that error by analyzing the worst-case situation, i.e. an illumination environment that implies the largest possible error.

To identify the worst case situation, we make an important assumption that the environment contributing indirect illumination does not contain any concentrated sources of illumination. Since the light emitters (i.e. luminaires) are not accounted for in the indirect calculation, this assumption is valid in most cases. There are some exceptions, though. An important example of concentrated illumination are caustics. Fortunately, caustics can be factored out from the indirect illumination computation by the use of photon mapping (see Section 4.3.3, pg. 65). Another example of concentrated sources of indirect illumination are patches of light filtering to an otherwise dark room through a window. This case (among others) breaks our assumption but Section 2.2.2.2 shows that an estimate of the actual illumination gradient at a point can be used to rectify problems arising when our assumption is not valid.

Assuming no concentrated sources, the worst case illumination is given by the *split sphere model* depicted in Figure C.1. The illumination environment is a sphere of radius R. Half of the sphere emits

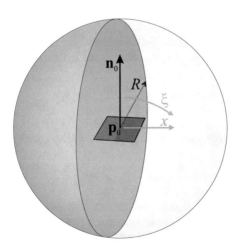

Figure C.1: The split sphere model. Illumination is coming from a sphere of radius R; half of the sphere emits zero radiance (black) and the other half non-zero radiance (white). A surface element is placed at the sphere center. The split sphere model represents an environment causing the fastest change of irradiance with translation and rotation of the surface element. Therefore, reusing the irradiance E_0 from the sphere center at a different location causes the largest possible error.

zero radiance (black) and the other half some non-zero constant radiance (white). A surface element, illuminated by the environment, is placed at the sphere center.

The split sphere model represents an environment causing the fastest change of irradiance with translation and rotation of the surface element and thus results in the largest possible interpolation error in an environment without concentrated sources of indirect illumination.

We want to estimate the error caused by reusing the irradiance E_0 from the split sphere center at another position with a different surface normal. Looking at Figure C.1, we see that the largest error with translation arises when the surface element is shifted along the direction perpendicular to the surface normal and to the boundary between the white and black halves of the split sphere. Rotation of the surface element around the axis lying in the split sphere boundary and perpendicular to the surface normal causes the largest error with rotation.

Reusing the original irradiance at a different location corresponds to the zeroth Taylor expansion of irradiance field as a function of translation distance and rotation angle. According to the Taylor's theorem, the error of this approximation is bounded (up to a constant) by the first derivative (i.e. the gradient) of the irradiance field:

$$\epsilon \leq \left| \frac{\partial E}{\partial x}(x - x_0) + \frac{\partial E}{\partial \xi}(\xi - \xi_0) \right|, \tag{C.1}$$

where x and ξ denote the translation distance and rotation angle, respectively.

To estimate the two derivatives, $\partial E/\partial x$ and $\partial E/\partial \xi$, we take advantage of the fact that irradiance is equal to the projection of the bright part of the split sphere to the base plane. Estimating the partial change of the projection area in terms of x gives us

$$\partial E = E_0 \frac{2R\partial x}{\frac{1}{2}\pi R^2} = E_0 \frac{4\partial x}{\pi R}.$$

Here, the denominator, $\frac{1}{2}\pi R^2$, is the area of a half circle and the numerator, $2R\partial x$, is the area of a stripe ∂x in width and $2R$ in length. In terms of ξ, the differential in the projection area is simply

$$\partial E = E_0 \frac{\frac{1}{2}\pi R^2 \partial \xi}{\frac{1}{2}\pi R^2} = E_0 \partial \xi.$$

Combining the irradiance derivatives with the triangle inequality gives us the error bound for the split sphere as:

$$\epsilon \leq \frac{4}{\pi} \frac{E_0}{R}|x - x_0| + E_0|\xi - \xi_0|.$$

The important result to notice here is that the error with respect to translation depends on the sphere radius R (or more generally on the distance to the geometry contributing indirect illumination), whereas the error with rotation is independent of the geometry.

Since we assume that no environment behaves worse in terms of error that the split sphere, we can generalize the above result to estimate the relative change in irradiance for any geometry. We replace $x - x_0$ with a distance between two points and $\xi - \xi_0$ with the angle between two surface normals (estimated, in our first-order approximation, as the sine between the two normals):

$$\epsilon(\mathbf{p}) \leq E_0 \left(\frac{4}{\pi} \frac{\|\mathbf{p} - \mathbf{p}_0\|}{R_0} + \sqrt{2 - 2\,\mathbf{n} \cdot \mathbf{n}_0} \right), \tag{C.2}$$

where $\mathbf{p}_0$ is the surface element location, $\mathbf{n}_0$ is the surface normal at $\mathbf{p}_0$, E_0 is the original irradiance at $\mathbf{p}_0$, and $\mathbf{p}, \mathbf{n}$ are the location and normal of the point where the irradiance is reused. Finally, R_0 is the harmonic mean distance to surfaces visible from point $\mathbf{p}_0$. The *harmonic* mean is used because the value of R_0 appears in the denominator, so the harmonic mean distributes the contributions to the overall error estimate correctly.

The derived error bound is used for two purposes. First, it serves us to decide which cached values can be used for interpolation at a point so that an error threshold is not exceeded. Second, the inverse of the error bound, divested of the numerical constants, $4/\pi$ and $\sqrt{2}$, which are specific to the split sphere model, is used as the weight in the irradiance interpolation procedure described in Section 2.2.1 on page 27.

APPENDIX D

Annotated References

D.1 PRIMARY SOURCES

The following is a list of papers that served as a primary source of information for this book.

[WRC88]   Gregory J. Ward, Francis M. Rubinstein, and Robert D. Clear. A Ray Tracing Solution for Diffuse Interreflection. In *Proceedings of SIGGRAPH*, pages 85–92, 1988.

The first paper on irradiance caching. Describes the overall algorithm (the "lazy evaluation" scheme), proposes the split sphere model to derive the weight function and record spacing, and suggests the use of an octree to index the cache records.
Covered in Chapter 2 and Section 4.3.2.

[WH92]   Gregory J. Ward and Paul Heckbert. Irradiance Gradients. In *Proceedings of Eurographics Workshop on Rendering*, pages 85–98, Bristol, UK, May 1992.

This paper introduces the translation and rotational irradiance gradients. It describes how the gradients can be estimated from the stratified hemisphere samples and how they are applied in interpolation to obtain smoother indirect illumination.
Covered in Section 2.1.1. (Note that the translation gradient formulation presented in Section 2.1.1 is based on [KGBP05] rather than on Ward and Heckbert's paper.)

[WLS98]   Greg Ward-Larson and Rob Shakespeare. Indirect Calculation. Chapter 12 in *Rendering with Radiance, The Art and Science of Lighting Visualization*. Morgan Kaufmann Publishers, 1998.

This book chapter summarizes the contents for the previous two papers with the focus on irradiance caching implementation in the Radiance lighting simulation system. The chapter also gives a summary of Radiance's parameters for irradiance caching.
Covered in Chapter 2 and Section 4.3.2.

[TL04]   Eric Tabellion and Arnauld Lamorlette. An approximate global illumination system for computer-generated films. In *Proceedings of SIGGRAPH*, 2004.

The paper gives a high-level description of tools, including irradiance caching, used at PDI/Dreamworks to support global illumination computation. Various modifications of the original irradiance caching algorithm are described. Integration of irradiance caching in a proprietary relighting tool is also sketched.
Covered throughout Chapters 2, 3, and 4.

[KGBP05]   Jaroslav Křivánek, Pascal Gautron, Kadi Bouatouch, and Sumanta Pattanaik. Improved radiance gradient computation. In *Proceedings of SCCG*, pages 149–153, 2005.

Describes a generalization of Ward and Heckbert's gradient computation [WH92] that can be used for radiance caching [KGPB05]. Specifically, Ward and Heckbert's assumption of cosine-proportional hemisphere sampling is lifted. The technique gives better results than the gradient computation described by the original radiance caching paper [KGPB05] since visibility changes are taken into account.
Covered in Section 2.1.1.

[KBPŽ06]   Jaroslav Křivánek, Kadi Bouatouch, Sumanta Pattanaik, and Jiří Žára. Making Radiance and Irradiance Caching Practical: Adaptive Caching and Neighbor Clamping. In *Rendering Techniques, Proceedings Eurographics Symposium on Rendering*, 2006.

Two independent contributions are described. First a method to adapt record density to actual illumination conditions. (This technique is somewhat similar to the irradiance caching implementation in Pixar's PrMan described by Christensen in [KGW+08], with the difference that it works in screen space rather than parametric surface space.) The second contribution is the neighbor clamping heuristic used to improve robustness of irradiance caching.
Neighbor clamping heuristic described in Section 2.2.2.4. Adaptive caching not covered in the book.

[GKBP05]   Pascal Gautron, Jaroslav Křivánek, Kadi Bouatouch, and Sumanta Pattanaik. Radiance cache splatting: A GPU-friendly global illumination algorithm. In *Proceedings of Eurographics Symposium on Rendering*, June 2005.

The paper reformulates irradiance caching to make it amenable to GPU implementation. Octree lookups are replaced by splatting the record contributions to the screen. GPU rasterization is used instead of hemisphere sampling. The algorithm gives up to $40\times$ speedup compared to Radiance *and affords for interactive walk-throughs with global illumination.*
Chapter 5 is an extended version of this paper.

[GBP07]   Pascal Gautron, Kadi Bouatouch, and Sumanta Pattanaik. Temporal radiance caching. *IEEE Transactions on Visualization and Computer Graphics*, 13(5), 2007.

This paper deals with the problem of illumination flickering animations. In addition to spatial interpolation, irradiance is interpolated in time using temporal gradients. This results in a flicker-free animation computed in shorter time than if the frames were computed independently.
Chapter 6 is an extended version of this paper.

[KGW+08]   Jaroslav Křivánek, Pascal Gautron, Greg Ward, Henrik Wann Jensen, Eric Tabellion, and Per Christensen. Practical global illumination with irradiance caching. In *ACM SIGGRAPH 2008 Classes*, 2008.

This class is the basis of the book. However, some material in the class is not covered in the book. The class materials are available from
`www.graphics.cornell.edu/~jaroslav/papers/2008-irradiance_caching_class/`.

D.2   FURTHER READING ON IRRADIANCE CACHING

The following papers are closely related to irradiance caching but are not covered in the book.

[KGPB05]   Jaroslav Křivánek, Pascal Gautron, Sumanta Pattanaik, and Kadi Bouatouch. Radiance caching for efficient global illumination computation. *IEEE Transactions on Visualization and Computer Graphics*, 11(5), September/October 2005.

Introduces radiance caching, an extension of irradiance caching that supports illumination interpolation on glossy surfaces. Spherical harmonics are used to represent the directional distribution of incoming radiance at a point. A novel gradient computation method is proposed.
Paper is briefly described in Section 4.2.4.2.

[AFO05]   Okan Arikan, David Forsyth, and James O'Brien. Fast and Detailed Approximate Global Illumination by Irradiance Decomposition. In *Proceedings of SIGGRAPH*, 2005.

This paper improves the performance of irradiance caching by reducing the number of cache records (and consequently the number of traced rays). The main idea is to divide indirect illumination into near and far field. Far field is computed with classical hemisphere sampling but the cache records can be more sparsely spaced (because the illumination is coming from distant surfaces). Illumination due to the near field is approximated by ignoring local visibility, which makes the computation much faster.
Not covered in the book.

[JDZJ08]   Wojciech Jarosz, Craig Donner, Matthias Zwicker, and Henrik Wann Jensen. Radiance caching for participating media. *ACM Trans. Graph.* 27(1), March 2008.

In this paper, irradiance and radiance caching are applied to accelerate global illumination computation in participating media. Gradients for single and multiple scattering terms are derived. The algorithm results in superior quality than photon mapping and faster rendering times than path tracing.
Not covered in the book.

[JZJ08]   Wojciech Jarosz, Matthias Zwicker, and Henrik Wann Jensen. Irradiance gradients in the presence of participating media and occlusions. *Computer Graphics Forum (Proceedings of EGSR 2008)*, 27(4), 2008.

This paper shows that some of the assumptions of the common irradiance gradient computation techniques are not valid in scenes with participating media. An irradiance gradient calculation algorithm is presented that takes into account the participating media and yields smoother interpolation. In addition to the gradients derived in Jarosz et al.'s TOG paper [JDZJ08], the gradient computation presented here takes visibility changes into account, thereby improving interpolation quality.
Not covered in the book.

D.3   BACKGROUND ON GLOBAL ILLUMINATION

[PH04]   Matt Pharr and Greg Humphreys. *Physically Based Rendering: from Theory to Implementation.* Morgan Kaufmann, 2004.

An excellent introductory book for realistic image synthesis. The book is a complete guide to building a physically-based ray tracer from scratch. It describes the background theory and also delves into the nitty-gritty implementation details.

[DBB06]   Philip Dutré, Kavita Bala, and Philippe Bekaert. *Advanced Global Illumination.* AK Peters, second edition, 2006.

This textbook focuses on the underlying concepts of global illumination computation from a more theoretical point of view than the PBRT book [PH04]. Some of the more recent practical global illumination algorithms are briefly described.

[Jen01]   Henrik Wann Jensen. *Realistic Image Synthesis Using Photon Mapping.* AK Peters, 2001.

This book provides a detailed description of the photon mapping algorithm for global illumination computation. Photon mapping and irradiance caching have complementary advantages and both algorithms benefit from their combination, as described in Section 4.3.3, pg. 65.

Bibliography

[AFO05]   Okan Arikan, David A. Forsyth, and James F. O'Brien. Fast and detailed approximate global illumination by irradiance decomposition. *ACM Trans. Graph. (SIGGRAPH 2005 Proceedings)*, 24(3), 2005. DOI: 10.1145/1073204.1073319

[AKDS04]   Thomas Annen, Jan Kautz, Frédo Durand, and Hans-Peter Seidel. Spherical harmonic gradients for mid-range illumination. *Proceedings of the Eurographics Symposium on Rendering 2004*, pages 331–336. Eurographics Association, 2004.

[BP04]   Michael Bunnell and Fabio Pellacini. *GPU Gems*, Addison Wesley, 2004.

[CB04]   Per H. Christensen and Dana Batali. An irradiance atlas for global illumination in complex production scenes. In *Rendering Techniques 2004 (Proceedings of the Eurographics Symposium on Rendering)*, pages 133–141, 2004.

[CCC87]   Robert L. Cook, Loren Carpenter, and Edwin Catmull. The Reyes image rendering architecture. In *SIGGRAPH '87*, pages 95–102, New York, NY, USA, 1987. DOI: 10.1145/37401.37414

[CG85]   Michael Cohen and Donald P. Greenberg. The Hemi-Cube: A Radiosity Solution for Complex Environments. In *Proceedings of SIGGRAPH*, pages 31–40, 1985. DOI: 10.1145/325165.325171

[Chr00]   Per H. Christensen. Faster photon map global illumination. *Journal of Graphics Tools*, 4(3):1–10, 2000.

[Chr03]   Per H. Christensen. Global illumination and all that. *RenderMan, Theory and Practice, SIGGRAPH '03 Course #9*, 2003.

[Chr05]   Per H. Christensen. The brick map as geometric primitive. Pixar, Emeryville, CA, 2005.

[CLF+03]   Per H. Christensen, David M. Laur, Julian Fong, Wayne L. Wooten, and Dana Batali. Ray differentials and multiresolution geometry caching for distribution ray tracing in complex scenes. *Computer Graphics Forum (Eurographics 2003 Conference Proceedings)*, 22(3):543–552, 2003.

[Coo89]   Robert L. Cook. *An Introduction to Ray tracing*. Morgan Kaufmann, 1989.

[CPC84]   Robert L. Cook, Thomas Porter, and Loren Carpenter. Distributed ray tracing. *SIGGRAPH Comput. Graph.*, 18(3):137–145, 1984. DOI: 10.1145/964965.808590

[CW93]   Michael F. Cohen and John R. Wallace. *Radiosity and Realistic Image Synthesis*. Morgan Kaufmann, 1993.

[DBB06]   Philip Dutré, Kavita Bala, and Philippe Bekaert. *Advanced Global Illumination*. AK Peters, second edition, 2006.

[Dut03]   Philip Dutré. Global illumination compendium. http://www.cs.kuleuven.ac.be/~phil/GI/, 2003.

[GBP07]   Pascal Gautron, Kadi Bouatouch, and Sumanta Pattanaik. Temporal radiance caching. *IEEE Transactions on Visualization and Computer Graphics*, 13(5), 2007. DOI: 10.1109/TVCG.2007.1061

[GKBP05]   Pascal Gautron, Jaroslav Křivánek, Kadi Bouatouch, and Sumanta Pattanaik. Radiance cache splatting: A GPU-friendly global illumination algorithm. In *Proceedings of Eurographics Symposium on Rendering*, 2005. DOI: 10.1145/1401132.1401231

[Gla89]   Andrew S. Glassner. *An Introduction to Ray tracing*. Morgan Kaufmann, 1989.

[Gla95]   Andrew S. Glassner. *Principles of Digital Image Synthesis*. Morgan Kaufmann, San Francisco, CA, 1995.

[Hec90]   Paul S. Heckbert. Adaptive radiosity textures for bidirectional ray tracing. In *Proceedings of SIGGRAPH*, 1990. DOI: 10.1145/97880.97895

[Hei91]   Tim Heidmann. Real shadows, real time. *Iris Universe*, 18:23–31, 1991.

[Jar08]   Wojciech Jarosz. *Efficient Monte Carlo Methods for Light Transport in Scattering Media*. PhD thesis, UC San Diego, September 2008.

[JDZJ08]   Wojciech Jarosz, Craig Donner, Matthias Zwicker, and Henrik Wann Jensen. Radiance caching for participating media. *ACM Trans. Graph.*, 27(1), 2008. DOI: 10.1145/1330511.1330518

[Jen01]   Henrik Wann Jensen. *Realistic Image Synthesis Using Photon Mapping*. AK Peters, 2001.

[JZJ08]   Wojciech Jarosz, Matthias Zwicker, and Henrik Wann Jensen. Irradiance gradients in the presence of participating media and occlusions. *Computer Graphics Forum (Proceedings of Eurographics Symposium on Rendering 2008)*, 27(4), 2008.

[Kaj86]   James T. Kajiya. The Rendering Equation. In *Proceedings of SIGGRAPH*, pages 143–150, 1986. DOI: 10.1145/15886.15902

[KBPv06]   Jaroslav Křivánek, Kadi Bouatouch, Sumanta N. Pattanaik, and Jiří Žára. Making radiance and irradiance caching practical: Adaptive caching and neighbor clamping. In *Rendering Techniques 2006, Eurographics Symposium on Rendering*. Eurographics Association, 2006. DOI: 10.1145/1401132.1401230

[Kel03]   Alexander Keller. Strictly deterministic sampling methods in computer graphics. In *Monte Carlo Ray Tracing, SIGGRAPH '03 Course #44*, 2003.

[KGBP05]   Jaroslav Křivánek, Pascal Gautron, Kadi Bouatouch, and Sumanta Pattanaik. Improved radiance gradient computation. In *Proceedings of SCCG*, pages 149–153, 2005. DOI: 10.1145/1090122.1090148

[KGPB05]   Jaroslav Křivánek, Pascal Gautron, Sumanta Pattanaik, and Kadi Bouatouch. Radiance caching for efficient global illumination computation. *IEEE Transactions on Visualization and Computer Graphics*, 11(5), 2005. DOI: 10.1109/TVCG.2005.83

[KGW⁺08]   Jaroslav Křivánek, Pascal Gautron, Greg Ward, Henrik Wann Jensen, Eric Tabellion, and Per Christensen. Practical global illumination with irradiance caching. In *ACM SIGGRAPH 2008 Classes*, 2008. DOI: 10.1145/1401132.1401213

[Kri05]   Jaroslav Krivánek. *Radiance Caching for Global Illumination Computation on Glossy Surfaces*. Phd thesis, Université de Rennes 1 and Czech Technical University in Prague, 2005.

[LAML07]   Thomas Larsson, Tomas Akenine-Möller, and Eric Lengyel. On faster sphere-box overlap testing. *Journal of graphics tools*, 12(1):3–8, 2007.

[Lan02]   Hayden Landis. Production-ready global illumination. In *RenderMan in Production, SIGGRAPH '04 Course #16*, 2002.

[LC04]   Bent Dalgaard Larsen and Niels Jorgen Christensen. Simulating photon mapping for real-time applications. In A. Keller and H. W. Jensen, editors, *Proceedings of Eurographics Symposium on Rendering*, pages 123–131, 2004.

[MFS04]   A. Méndez-Feliu and Mateu Sbert. Comparing hemisphere sampling techniques for obscurance computation. In *International Conference on Computer Graphics and Artificial Intelligence 3IA 2004*, 2004.

[NK04]   Ivan Neulander and Tae-Yong Kim. Lighting hair with high dynamic range images. In *Photorealistic Hair Modeling, Animation, and Rendering, SIGGRAPH '04 Course #9*, 2004.

[PH04]   Matt Pharr and Greg Humphreys. *Physically Based Rendering: from Theory to Implementation*. Morgan Kaufmann, 2004.

[Pho75]   Bui Tuong Phong. Illumination for computer generated pictures. *Communications of the ACM*, 18(6):311–317, 1975. DOI: 10.1145/360825.360839

[Ros06]   Sheldon M. Ross. *Introduction to Probability Models*. Academic Press, Inc., Orlando, FL, USA, 2006.

[RSC87]   William T. Reeves, David H. Salesin, and Robert L. Cook. Rendering antialiased shadows with depth maps. In *Proceedings of SIGGRAPH*, volume 21, pages 283–291, 1987. DOI: 10.1145/37402.37435

[SA08]   Mark Segal and Kurt Akeley. *The OpenGL Graphics System: A Specification (Version 3.0 - August 11, 2008)*. 2008.

[SD02]   Marc Stamminger and George Drettakis. Perspective shadow maps. In *Proceedings of SIGGRAPH*, 2002.

[SKDM05]   Miloslaw Smyk, Shin-ichi Kinuwaki, Roman Durikovic, and Karol Myszkowski. Temporally coherent irradiance caching for high quality animation rendering. *Computer Graphics Forum (Proceedings of Eurographics)*, 24(3):401–412, 2005. DOI: 10.1111/j.1467-8659.2005.00865.x

[SP89]   Francois Sillion and Claude Puech. A General Two-Pass Method Integrating Specular and Diffuse Reflection. In *Proceedings of SIGGRAPH*, pages 335–344, 1989. DOI: 10.1145/74334.74368

[SWZ96]   Peter Shirley, Changyaw Wang, and Kurt Zimmerman. Monte Carlo techniques for direct lighting calculations. *ACM Transactions on Graphics*, 15(1), 1996. DOI: 10.1145/226150.226151

[TL04]   Eric Tabellion and Arnauld Lamorlette. An approximate global illumination system for computer-generated films. In *Proceedings of SIGGRAPH*, 2004. DOI: 10.1145/1186562.1015748

[TMS02]   Takehiro Tawara, Karol Myszkowski, and Hans-Peter Seidel. Localizing the final gathering for dynamic scenes using the photon map. In *VMV*, 2002.

[Ups90]   Steve Upstill. *The RenderMan Companion: A Programmer's Guide to Realistic Computer Graphics*. Addison-Wesley Professional, 1990.

[Vea97]   Eric Veach. *Robust Monte Carlo Methods for Light Transport Simulation*. PhD thesis, Stanford University, 1997.

[War92]   Gregory J. Ward. Measuring and modeling anisotropic reflection. In *Proceedings of SIGGRAPH*, 1992. DOI: 10.1145/142920.134078

[War94]   Gregory J. Ward. The RADIANCE Lighting Simulation and Rendering System. In *Computer Graphics Proceedings, Annual Conference Series, 1994 (Proceedings of SIGGRAPH*, pages 459–472, 1994. DOI: 10.1145/192161.192286

[WDP99]   Bruce Walter, George Drettakis, and Steven Parker. Interactive rendering using the render cache. In *Proceedings of Eurographics Workshop on Rendering*, pages 235–246, 1999.

[WH92]   Gregory J. Ward and Paul Heckbert. Irradiance Gradients. In *Proceedings of Eurographics Workshop on Rendering*, pages 85–98, Bristol, UK, 1992.

[Wil78]   Lance Williams. Casting curved shadows on curved surfaces. In *Proceedings of SIGGRAPH*, pages 270–274, 1978. DOI: 10.1145/965139.807402

[WLS98]   Greg Ward-Larson and Rob Shakespeare. *Rendering with Radiance, The Art and Science of Lighting Visualization*. Morgan Kaufmann Publishers, 1998.

[WRC88]   Gregory J. Ward, Francis M. Rubinstein, and Robert D. Clear. A Ray Tracing Solution for Diffuse Interreflection. In *Proceedings of SIGGRAPH*, pages 85–92, 1988. DOI: 10.1145/378456.378490

About the Authors

JAROSLAV KŘIVÁNEK

Jaroslav Křivánek is a visiting researcher at the Cornell University Program of Computer Graphics. Prior to this appointment, he worked as assistant professor at the Czech Technical University in Prague. He received his Ph.D. from IRISA/INRIA Rennes and Czech Technical University (joint degree) in 2005. In 2003 and 2004 he was a research associate at the University of Central Florida. He received a Masters in computer science from the Czech Technical University in Prague in 2001. Jaroslav's research focuses mainly on realistic rendering and global illumination.

PASCAL GAUTRON

Pascal Gautron is a R&D Engineer at Thomson Corporate Research France. In 2006 he received a Ph.D. in computer graphics from the University of Rennes 1, France, in collaboration with the University of Central Florida. He was a postdoctoral fellow at Orange Labs in 2007. His main research interest is high quality real-time rendering using graphics hardware, including global illumination and participating media.

Index

Printed in the United States
by Baker & Taylor Publisher Services